Wearable and Implantable Medical Devices

Applications and Challenges

Advances In Ubiquitous Sensing Applications for Healthcare

Wearable and Implantable Medical Devices

Applications and Challenges

Volume Seven

Series Editors

Nilanjan Dey

Amira S. Ashour

Simon James Fong

Volume Editors

Nilanjan Dey
Techno India College of Technology, West Bengal, India

Amira S. Ashour
Tanta University, Egypt

Simon James Fong
University of Macau, Av. Padre Tomas Pereira, Taipa, Macau SAR

Chintan Bhatt
Charotar University of Science and Technology, Anand, India

ELSEVIER

ACADEMIC PRESS
An imprint of Elsevier

Academic Press is an imprint of Elsevier
125 London Wall, London EC2Y 5AS, United Kingdom
525 B Street, Suite 1650, San Diego, CA 92101, United States
50 Hampshire Street, 5th Floor, Cambridge, MA 02139, United States
The Boulevard, Langford Lane, Kidlington, Oxford OX5 1GB, United Kingdom

British Library Cataloguing-in-Publication Data
A catalogue record for this book is available from the British Library

Library of Congress Cataloging-in-Publication Data
A catalog record for this book is available from the Library of Congress

ISBN: 978-0-12-815369-7

For Information on all Academic Press publications
visit our website at https://www.elsevier.com/books-and-journals

Publisher: Mara Conner
Acquisition Editor: Fiona Geraghty
Editorial Project Manager: John Leonard
Production Project Manager: Kamesh Ramajogi
Cover Designer: Victoria Pearson

Typeset by MPS Limited, Chennai, India

Contents

List of contributors

Jon Bilbatua Andrés
Markontzaga Health Center, Osakidetza-Basque Health System, Sestao, Spain

Farah Bader
Centre for Sustainable Manufacturing and Recycling Technologies (SMART), The Wolfson School of Mechanical, Electrical and Manufacturing Engineering, Loughborough University, Leicestershire, United Kingdom

Sarra Berrahal
Communications Networks and Security Research Laboratory, University of Carthage, Tunisia

Chintan Bhatt
Charotar University of Science and Technology, Anand, India

N.P.G. Bhavani
Department of EEE, Meenakshi College of Engineering, Chennai, India

Noureddine Boudriga
Computer Science Department, University of Western Cape, Belleville, Cape Town, South Africa

Su-Qun Cao
Faculty of Mechanical and Materials Engineering, Huaiyin Institute of Technology, Huaian Shi, China

Nicholas Cummins
ZD.B Chair of Embedded Intelligence for Health Care and Wellness, University of Augsburg, Germany

Ankur Dumka
Graphic Era (Deemed to be University), Dehradun, India; University of Petroleum and Energy Studies, Dehradun, India

G. Durgadevi
Department of ECE, New Prince Shri Bhavani College of Engineering and Technology, Chennai, India

Pablo Fdez-Arroyabe
Geography Department, University of Cantabria, Cantabria, Spain

Diego Soliño Fernández
Lead Business Development at WearHealth UG, WearHealth, Spain

Konstantinos Fysarakis
Sphynx Technology Solutions AG, Zug, Switzerland

Anita Gehlot
School of Electronics & Electrical Engineering, Lovely Professional University, India

Rajeswari Hari
Department of IBT, Dr. MGR Educational and Research Institute, Chennai, India

Sotiris Ioannidis
Institute of Computer Science, Foundation for Research and Technology–Hellas, Greece

Sandeep Jagtap
Centre for Sustainable Manufacturing and Recycling Technologies (SMART), The Wolfson School of Mechanical, Electrical and Manufacturing Engineering, Loughborough University, Leicestershire, United Kingdom; The National Centre for Food Manufacturing, University of Lincoln, Holbeach, United Kingdom

Vivaksha J. Jariwala
Sarvajanik College of Engineering and Technology, Surat, India

Devesh C. Jinwala
S.V. National Institute of Technology, Surat, India

V. Karthikeyan
Department of EEE, Dr. MGR Educational and Research Institute, Chennai, India

Jatin Kumar Khilrani
EE, American Megatrends India Pvt. Ltd., Tamil Nadu, India

Vidhi Kokel
Charotar University of Science and Technology, Anand, India

Pooja Kothari
Charotar University of Science and Technology, Anand, India

Dimitrios Koutsouris
Institute of Communication and Computer Systems, Athens, Greece

K. Senthil Kumar
Department of ECE, Dr. MGR Educational and Research Institute, Chennai, India

S.R. Liyanage
University of Kelaniya, Kelaniya, Sri Lanka

Manolis Lourakis
Institute of Computer Science, Foundation for Research and Technology–Hellas, Greece

Evangelos Loutsetis
Institute of Computer Science, Foundation for Research and Technology–Hellas, Greece

Michail Maniadakis
Institute of Computer Science, Foundation for Research and Technology–Hellas, Greece

Shweta Masrani
Charotar University of Science and Technology, Anand, India

Mamta Mittal
Department of Computer Science & Engineering, G.B. Pant Government Engineering College, Okhla, India

Maria Pateraki
Institute of Computer Science, Foundation for Research and Technology–Hellas, Greece

R.S. Ponmagal
Department of CSE, Dr. MGR Educational and Research Institute, Chennai, India

Anushree Sah
Graphic Era (Deemed to be University), Dehradun, India; University of Petroleum and Energy Studies, Dehradun, India

Vangelis Sakkalis
Institute of Computer Science, Foundation for Research and Technology–Hellas, Greece

Björn Schuller
ZD.B Chair of Embedded Intelligence for Health Care and Wellness, University of Augsburg, Germany

Rajesh Singh
School of Electronics & Electrical Engineering, Lovely Professional University, India

Georgios Spanoudakis
Sphynx Technology Solutions AG, Zug, Switzerland

V. Srividhya
Department of EEE, Meenakshi College of Engineering, Chennai, India

K. Sujatha
Department of EEE, Dr. MGR Educational and Research Institute, Chennai, India

Iraklis Varlamis
Institute of Computer Science, Foundation for Research and Technology–Hellas, Greece; Department of Informatics and Telematics, Harokopio University of Athens, Athens, Greece

Preface

Shortening the duration of hospital stays is needed to reduce healthcare costs among other things. Wearable devices and sensors could also help improve healthcare and reduce costs by providing constant health monitoring. Technological advancements decrease the monitoring gadgets and power sources, which opening up a totally different universe of conceivable healthcare out comes. The groundbreaking use of mobile phones facilitates real-time monitoring, and wearable and implantable technologies and devices allow sensing of parameters of various diseases. Such technologies can transfer data to a remote center to automatically perform functions based on the sensors' readings. Thus wearable and implantable medical devices have the potential to provide significant economic savings as well as to increase patients' quality of life.

This book includes 10 chapters with the following studies.

In Chapter 1, Internet of Things–triggered and power-efficient smart pedometer algorithm for intelligent wearable devices, Singh and Gehlot propose a novel intelligent wearable device with a pedometer algorithm to detect and record physical movements. This device is used to monitor workouts using a smart algorithm for step detection that analyzes the instantaneous acceleration versus time waveform and movements involved in stepping.

In Chapter 2, Biosensors and Internet of Things in smart healthcare applications: challenges and opportunities, Pateraki et al. highlight the challenges and opportunities of the application of smart biosensors in healthcare and present three state-of-the-art solutions leveraging smart sensors.

In Chapter 3, Wearable electroencephalography technologies for brain-computer interfacing using the P300-based BCI communication techniques was discussed for locked-in patients while summarizing the current state of wireless BCI systems. The hardware and software requirements that should be satisfied to enable ad-hoc brain–computer and brain–brain networks are identified. The state-of-the-art in wearable BCI systems are also reviewed and future possibilities in wireless BCI research explored.

Jariwala, in Chapter 4, AdaptableSDA: secure data aggregation framework in wireless body area networks, proposed a framework called the AdaptableSDA to provide essential security attributes to network nodes wired in a specific data aggregation topology. In AdaptableSDA, a privacy homomorphism is used to ensure data integrity and provide secure data aggregation in wireless body area networks.

Chapter 5, Screening and early identification of microcalcifications in breast using texture-based ANFIS classification, investigates robust detection schemes to detect breast cancer early. This noninvasive method uses texture features to diagnose the microcalcifications (cancerous lesions) in mammary glands at the premature stage of cancer.

Chapter 6, Work environment and healthcare: a biometeorological approach based on wearables, covers the potential of wearable devices and portable meteorological sensors in different scientific disciplines, particularly in biometeorology. The combination of physiological data from wearable devices and environmental data could help the scientific community identify worker vulnerability to heat stress at an aggregate and individual scale.

In Chapter 7, Reading Assistant: a reciter in your pocket, Bhatt et al. discuss a device called the Reading Assistant and its unique contribution to the field of wearable devices.

The main objective of Chapter 8, Toward secure and private WIBSN-based health monitoring applications, is to give a thorough analysis of the adoption of wearable devices and sensors in the healthcare sector by identifying possible security and privacy threats as well as their countermeasures.

Chapter 9, Smart ambulance traffic management system (SATMS)—a support for wearable and implantable medical devices, offers a traffic management solution for ambulances and other first responders based on the requirements of a particular area, city, district, or state.

Finally, Chapter 10, Internet of things-linked wearable devices for managing food safety in the healthcare sector, covers the role and benefits of implementing wearable technologies to collect crucial food product processing and development data for use for real-time food safety in healthcare organizations.

This volume of cutting-edge research delivers insights into the opportunities and novel strategies, techniques, and challenges for wearable and implantable medical devices and is a key resource for researchers, engineers, and developers.

The editors thank the authors for their high-quality contributions and also the reviewers for their accurate, detailed, and timely reviews. Special thanks also to our publisher, Elsevier. We hope this book will stimulate further research in wearable and implantable medical technologies.

Volume Editors

Nilanjan Dey, India

Amira S. Ashour, Egypt

Simon James Fong, China

Chintan Bhatt, India

CHAPTER

1 Internet of Things—triggered and power-efficient smart pedometer algorithm for intelligent wearable devices

Rajesh Singh[1], Anita Gehlot[1], Jatin Kumar Khilrani[2] and Mamta Mittal[3]

[1]*School of Electronics & Electrical Engineering, Lovely Professional University, India*

[2]*EE, American Megatrends India Pvt. Ltd., Tamil Nadu, India*

[3]*Department of Computer Science & Engineering, G.B. Pant Government Engineering College, Okhla, India*

Chapter Outline

1.1 Introduction

Wearable technology includes smart devices that can be worn on the body. Intelligent devices like activity trackers keep track of data without human interference. Examples of these types of devices include smartwatches and activity trackers such as Apple Watch Series 2 or the Samsung Galaxy Gear Sport. This type of wearable technology can also be used in health-monitoring devices.

Since the wearable technology available today is costly here we propose a cost-effective device that can not only can count footsteps but also communicate

Wearable and Implantable Medical Devices. DOI: https://doi.org/10.1016/B978-0-12-815369-7.00001-X

environmental parameters through a mobile app and Internet of Things (IoT) server. To maintain optimal health people should take 10,000 steps a day (equivalent to 5 miles per day) [1]. A pedometer is a device that measures steps taken by physical activity of the hands or hips [2]. Studies have shown that pedometers use results in a greater increase in leisure walking without any effect on overall activity level [3]. Measuring physical activity is a key part of studying health [4]. In fact, people used to wear bells on their hands to count steps. Of course, today, with the evolution of MEMS technology sensors are used to measure physical activity. Since in the wearable market, the major challenge is still battery life, this chapter presents a power-efficient pedometer algorithm for wearables.

The literature shows different approaches to calculating steps considering accuracy and power efficiency. For example, some researchers have studied the gyroscope [5,6], which suffers from drift and recalibration and is not power efficient. Some of the commercially available smartphone applications such as the "Walkmeter" [7] use GPS to track motion, but this device is also a power hungry and requires long bootup time. The paper "The Accuracy of Pedometers in Measuring Walking Steps on a Treadmill in College Students" [8] explains how step count varies in different fitness devices. This paper discussed an algorithm using a three-axis accelerometer, but only the axis with maximum change was studied [9], which is not applicable for all people. Accelerator is used as a source and neural network is used for processing of data [10].

Embedded wearable devices needs to balance the performance and power both. Jahan et al. described the relation between steps per day (using a pedometer), self-reported physical activity, and health indices [11]. De Craemer et al. compared omron walking pedometers and accelerometer-based step counters (actigraph) [12]. Feng et al. measured step count cut points (using Youden's J statistic) using actigraph accelerometers [13]. Feng et al. suggested an appropriate accelerometer to count steps at slower walking speeds and suggested ways to improve the algorithms of accelerometers [14]. Singh et al. concluded that the FitBit Zip has poor accuracy of 0.7 m/s or chest and hip [15].

The intelligent wearable device (IWD) pedometer algorithm discussed here is based on accelerometer sensor data that gives values on the 3D axis (i.e., X-axis, Y-axis, and Z-axis). The accelerometer value is the measure of acceleration experience to freefall [16]. The signals from the accelerometer consist of noise due to transducer elements with electrical and mechanical properties. This noise is comprised of Gaussian noise and unwanted vibration from daily activity. In practical scenarios, steps while driving also need to be removed with the help of orientation and dynamic filtration methods. The orientation of the accelerometer can be defined in mathematical terms such as yaw, pitch, and roll [17].

Toth et al. proposed a 14-step counting method to determine accuracy under open living conditions of the GoPro camera [18]. Alhusaini et al. conducted a study to compare the physical activities between children with Down's syndrome and healthy children [19]. Coelho et al. investigated pedometer use for asthma patients to measure physical activity [20]. Binh et al. proposed two nature-based algorithms—Improved Cuckoo Search and Chaotic Flower Pollination—to improve the performance of wireless sensor networks. The algorithm showed the better results in terms of computation time, stability, and solution quality [21]. Roy et al. discussed the role of cryptography in wireless sensor networks with cellular automata. A hybrid CA was framed for encryption and decryption of sensory data was used [22]. Bhatt et al. discussed the role of IoT in healthcare applications. The association of sensors and wireless techniques are discussed thoroughly [23].

1.2 Intelligent wearable device description

The IWD pedometer aims at efficient calculation of steps and calories from accelerometer signals after filtering out noise. It utilizes the low-power mode of the microcontroller and sensors to achieve a better power-efficient pedometer algorithm. Fig. 1.1 shows the generalized block diagram of the IWD network. Multiple IWD devices are proposed, which may be connected to a common server, to monitor physical activity at a centralized place. The controller consumes less power in idle state.

Fig. 1.2 shows the block diagram of IWD, including the microcontroller, accelerometer, and battery. Fig. 1.3A shows the block diagram of the IWD smart system, which is comprised of Arduino Nano, NodeMCU (WiFi device),

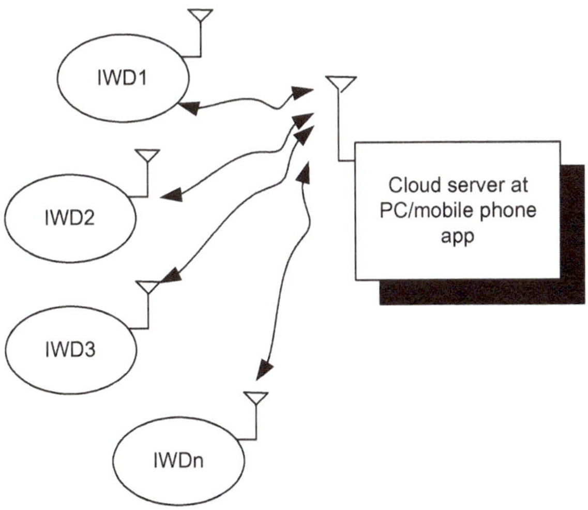

FIGURE 1.1

Generalized block diagram of intelligent wearable device (IWD) network.

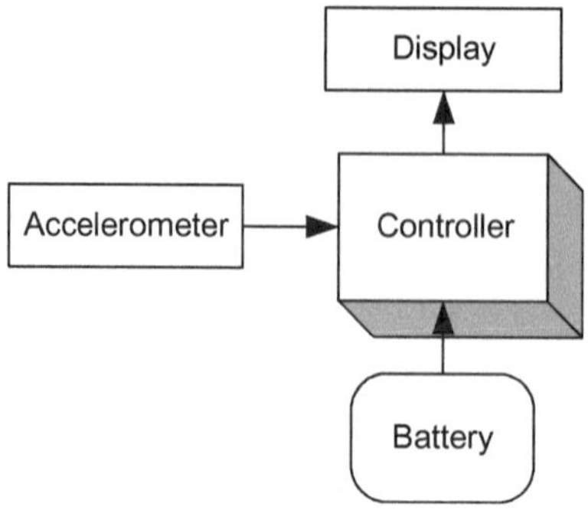

FIGURE 1.2

Block diagram of the intelligent wearable device.

Bluetooth and LCD20*4, and battery as power supply. NodeMCU communicates the data to the cloud server as well as to the app.

Fig. 1.3B shows the schematic diagram of the proposed IWD algorithm.

1.3 Intelligent wearable device pedometer algorithm and evaluation

Fig. 1.4 shows the flowchart of the IWD pedometer algorithm. It is based on accelerometer data that gives 3D axis signals, namely *X*-axis, *Y*-axis, and *Z*-axis. These coordinates can be used to monitor the physical movements of the person.

The vertical lines in Figs. 1.5–1.7 represents the point where peak changes are detected. The type of noise generated by the transducer is removed by applying sensitivity to the *X*-, *Y*-, and *X*-axes. As high-frequency vibrations suffer from Gaussian noise and low-frequency motion suffer from drift, so frequency needs to be set.

An optimal frequency is set as 100 Hz considering the capability of a typical user (e.g., eight steps per second). Accelerometer data is passed through the discrete moving average filter that is responsible for smoothing the Gaussian noise. The region needs to identified where the probability of steps is more (e.g., people used to walk generally putting the hand toward the ground). The pitch helps to define the zone of activity. Right-hand pitch is positive and left-hand pitch is negative. Eq. (1.1) explains pitch (Figs. 1.8–1.10).

$$\text{pitch} = \frac{180}{\text{pi}} {}^{*} atan\left[\frac{x}{z^2 + y^2}\right] \tag{1.1}$$

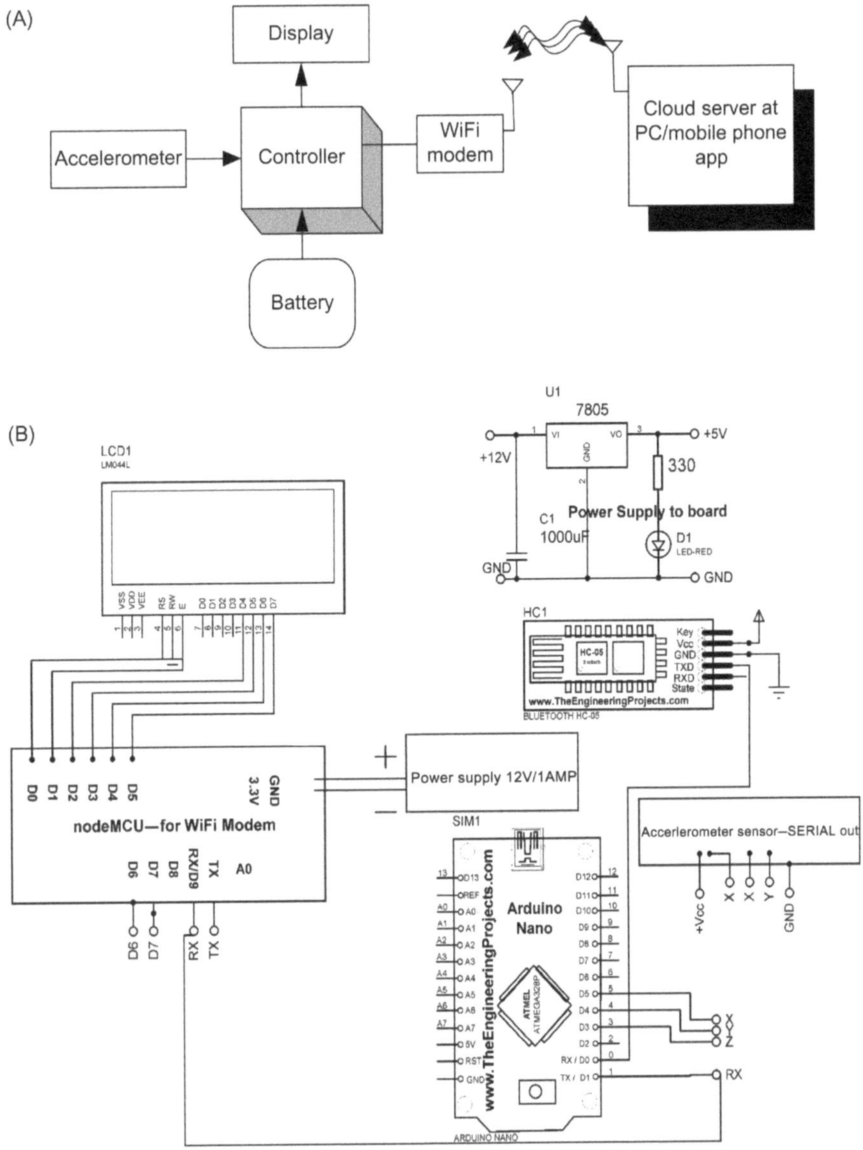

FIGURE 1.3

(A) Block diagram of the system. (B) Hardware schematic of the intelligent wearable device (IWD).

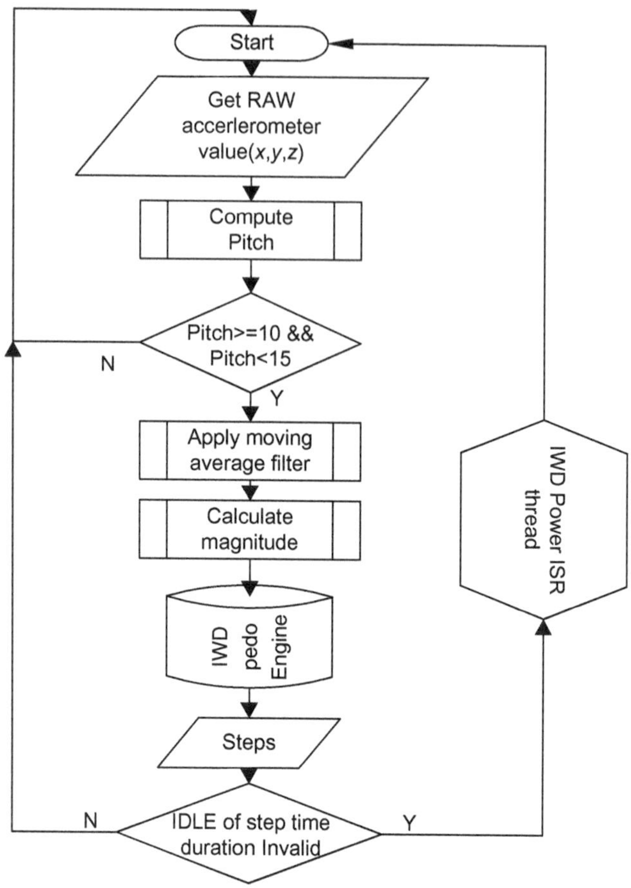

FIGURE 1.4

Algorithm flowchart.

1.4 Application development

1.4.1 ThingSpeak server

The ThingSpeak server can be used to collect the data for further analysis. It can be accessed from anywhere in the world and is always online (Figs. 1.11–1.13).

Use the following steps to create a channel:

1. Sign in to ThingSpeak by creating a new MathWorks account.
2. Click Channels >My Channels.
3. Click New Channel.
4. Check the boxes next to fields 1–1. Enter the channel setting values.
5. Click on Save Channel at the bottom of the settings.
6. Check API write key (this key needs to write in the program for local server).

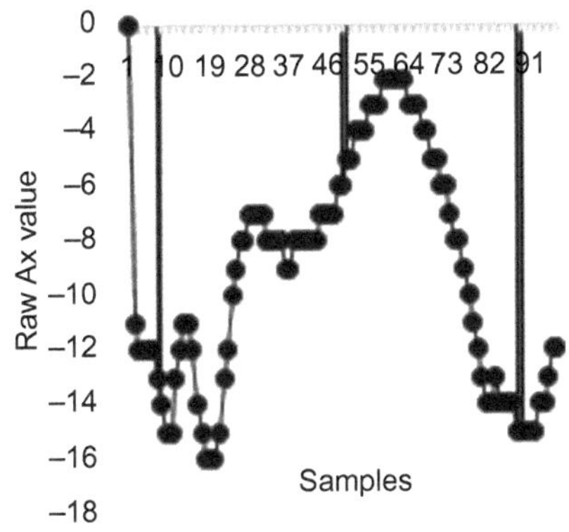

FIGURE 1.5

X-axis sample.

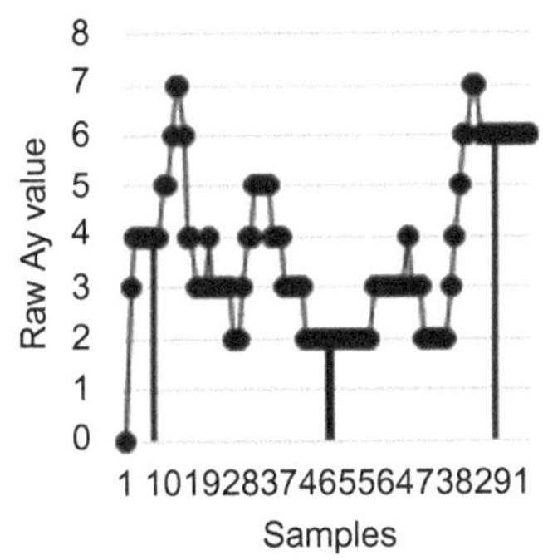

FIGURE 1.6

Y-axis sample.

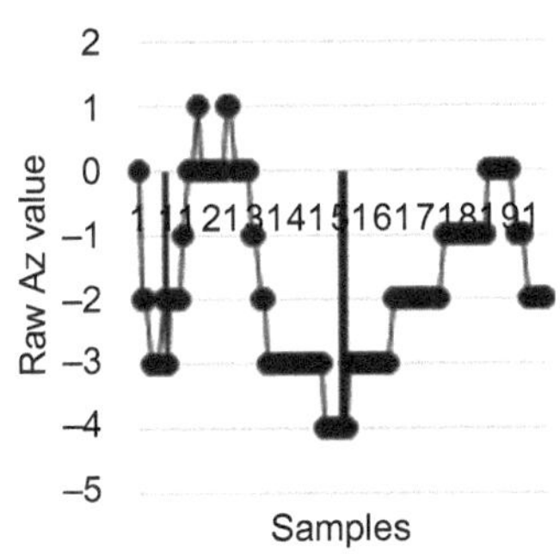

FIGURE 1.7

Z-axis sample.

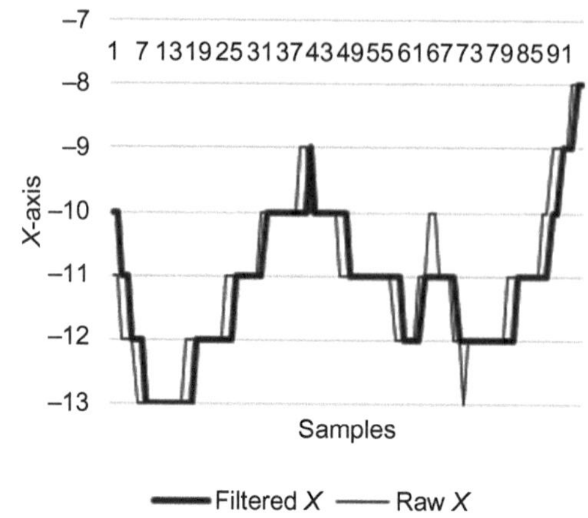

FIGURE 1.8

X-axis moving average filter.

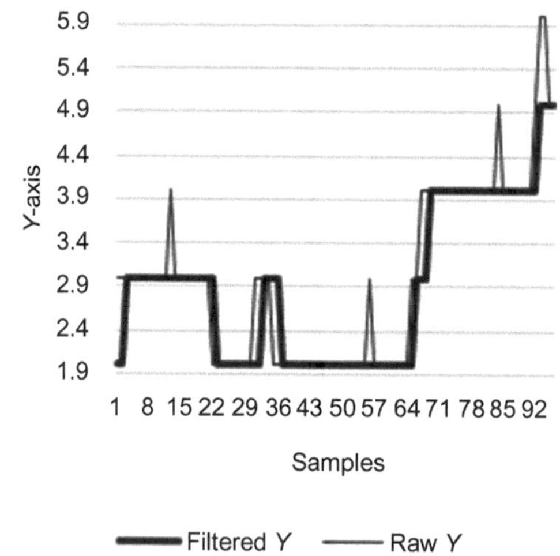

FIGURE 1.9

Y-axis moving average filter.

1.4.2 Virtuino app

The Virtuino app is a human machine interface platform on the cloud. With the help of this mobile app the sensory data can be flashed on the mobile also. Virtuino application is controlled through Bluetooth, WiFi, GPRS, and ThingSpeak. It is a freely available application from IliasLamprou.

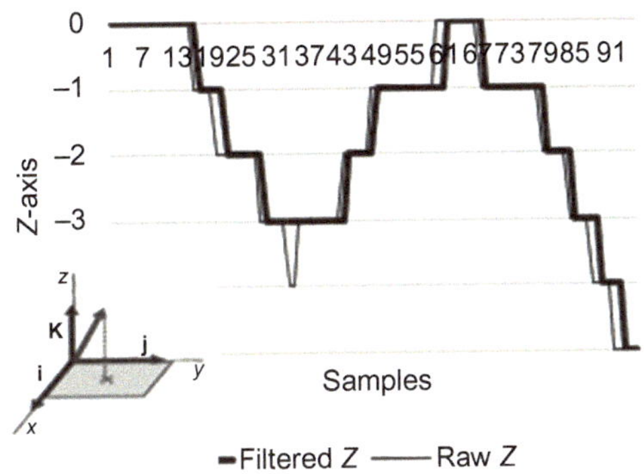

FIGURE 1.10

Z-axis moving average filter.

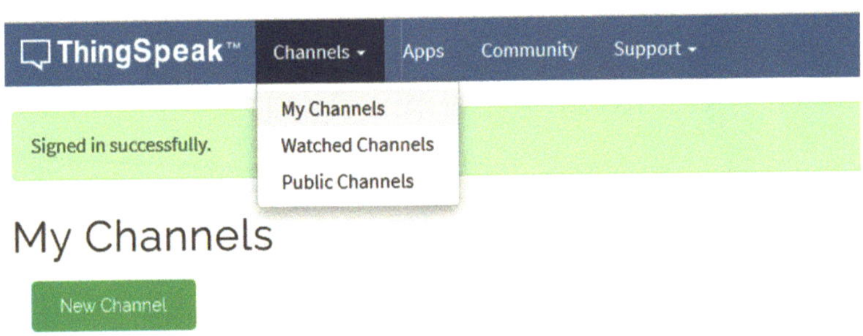

FIGURE 1.11

Window in ThingSpeak.

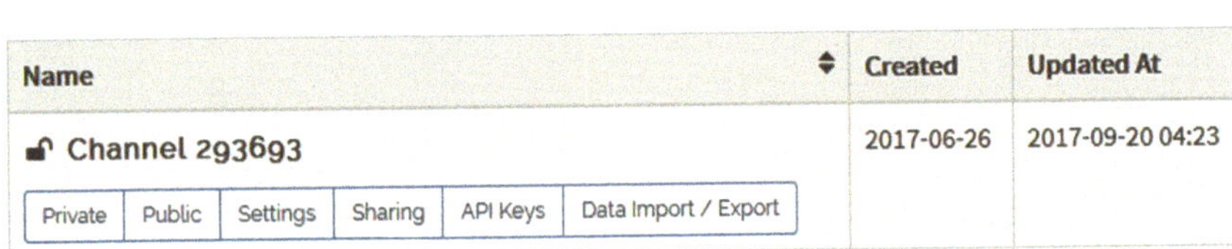

FIGURE 1.12

New Channel in My Channels.

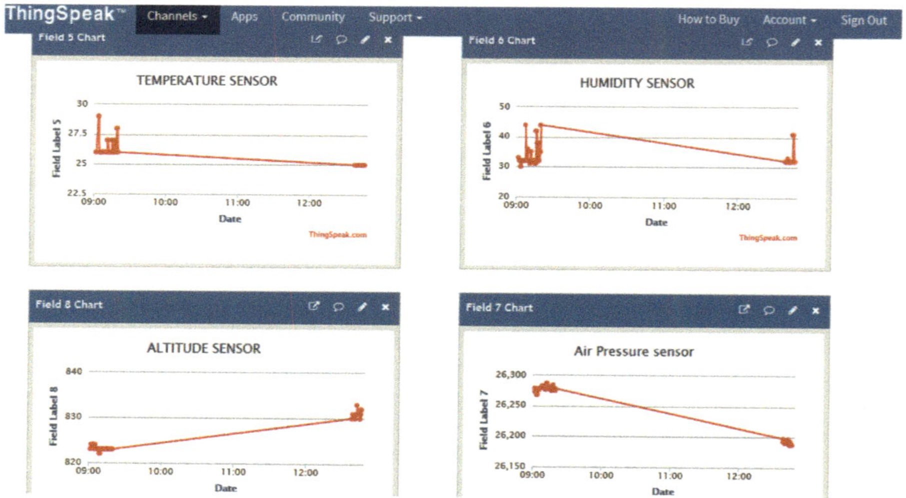

FIGURE 1.13

Charts from the fields showing data received from local server.

Step 1: make connection among Arduino Uno, NodeMCU, and external devices.
Step 2: Follow these steps:

a. Click to download the Virtuino Library ver 1.1.
b. Run to add the library via Arduino IDE.
c. Burn the hex code in NodeMCU.

Step 3: Add WiFi settings with Android device.
Step 4: Create the app and run it to interact with ESP8266/NodeMCU.

1.5 Software development

//// setup function for step counter in Virtuino app

The given function is used to initialize and setup( ) function in the program for the step counter in the Virtuino app.

```
void setup()
{
  virtuino.DEBUG=true;
  virtuino.password="1234";
  Serial.begin(9600);
  delay(10);

  pinMode(2, OUTPUT);
  digitalWrite(2, 0);

  Serial.println("Connecting to "+String(ssid));
  WiFi.mode(WIFI_STA);
  WiFi.begin(ssid, password);
   while (WiFi.status() != WL_CONNECTED)
   {
     delay(500);
     Serial.print(".");
   }
   Serial.println("");
   Serial.println("WiFi connected");
   Serial.println(WiFi.localIP());
  server.begin();
  Serial.println("Server started");
}
```

/////// loop function of step counter in Virtuino app

The given function is used to initialize and loop( ) function in the program for the step counter in the Virtuino app.

```
void loop()
{
virtuino.run();
int LEFT=virtuino.vDigitalMemoryRead(0);
serial_event();
Serial.print( AXIS_X);
Serial.print( AXIS_Y);
Serial.println( AXIS_Z);
   AXIS_X=97;
   AXIS_Y=76;
   AXIS_Z=38;

   virtuino.vMemoryWrite(5,AXIS_X);
   virtuino.vMemoryWrite(6,AXIS_Y);
   virtuino.vMemoryWrite(7,AXIS_Z);
   long t= millis();
    if (t>stored_Time+5000)
    {
      counter_virtuino++;
      if (counter_virtuino>20) counter_virtuino=0;
      stored_Time = t;
      virtuino.vMemoryWrite(12,counter_virtuino);
    }
  }
```

Serial event for the step counter in the Virtuino app.

The given function is used to initialize and serial_event( ) function in the program for the step counter in the Virtuino app.

```
void serial_event()
{
  if (Serial.available()<1)  return;
  char S=Serial.read();
  if (S!='\r')                 return;
   AXIS_X=Serial.parseInt();
   AXIS_Y=Serial.parseInt();
   AXIS_Z=Serial.parseInt();
}
```

/////// setup function of the step counter in ThingSpeak

The given function is used to initialize and setup( ) function in the program for step counter in ThingSpeak server.

```
void setup()
{
Serial.begin(9600);
delay(10);
WiFi.begin(ssid, password);
Serial.println();
Serial.println();
Serial.print("Connecting to ");
Serial.println(ssid);

WiFi.begin(ssid, password);

while (WiFi.status() != WL_CONNECTED)
{
delay(500);
Serial.print(".");
}
Serial.println("");
Serial.println("WiFi connected");
}
```

/////////// loop function for step counter for ThingSpeak

The given function is used to initialize and loop( ) function in the program for step counter in ThingSpeak server.

```
void loop()
{

        if (client.connect(server,80))
         {
          read_SENSOR_NODEMCU();
          send1_PEDOMETER_PARA();
         }
        client.stop();
        Serial.println("Waiting");
        delay(20000);

 }
```

/////// function to send data to cloud server

The given function is used to initialize and send1_PEDOMETER_PARA() function in the program for step counter in ThingSpeak server.

```
void send1_PEDOMETER_PARA()
{

        String postStr = apiKey1;
        postStr +="&field1=";
        postStr += String(SERIAL_DATA1);
        postStr +="&field2=";
        postStr += String(SERIAL_DATA2);
        postStr +="&field3=";
        postStr += String(SERIAL_DATA3);
        postStr += "\r\n\r\n";
        client.print("POST /update HTTP/1.1\n");
        client.print("Host: api.thingspeak.com\n");
        client.print("Connection: close\n");
        client.print("X-THINGSPEAKAPIKEY: "+apiKey1+"\n");
        client.print("Content-Type: application/x-www-form-urlencoded\n");
        client.print("Content-Length: ");
        client.print(postStr.length());
        client.print("\n\n");
        client.print(postStr);
        Serial.print("Send data to channel-1 ");
        Serial.print("Content-Length: ");
        Serial.print(postStr.length());
        Serial.print("Field-1: ");
        Serial.print(SERIAL_DATA1);
        Serial.print("Field-2: ");
        Serial.print(SERIAL_DATA2);
        Serial.print("Field-3: ");
        Serial.print(SERIAL_DATA3);
        Serial.println(" data send");
```

1.6 Results

A step is defined as combination of acceleration and deceleration. The unit cycle can be detected by taking the axis with the highest change, but this may be different for all people as everyone move hands differently while they walking and thus the dominancy changes as observed in Figs. 1.14 and 1.15. In most cases the x-axis has the highest order of change. The IWD pedometer algorithm considers the results of the i, j, and k vector forces called X-, Y-, and Z-axis magnitude of force, which is calculated using Eq. (1.2).

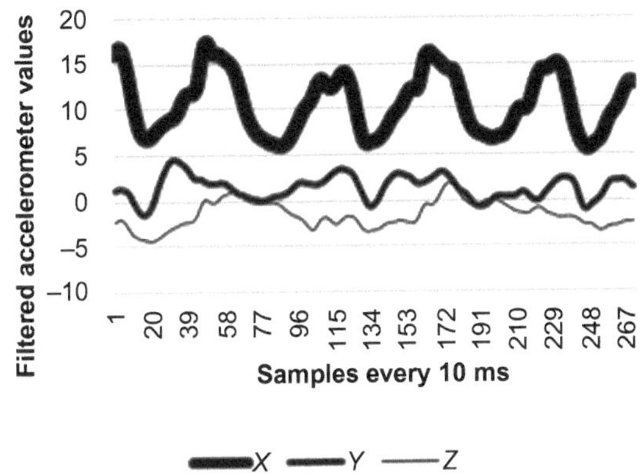

FIGURE 1.14

X-axis dominant.

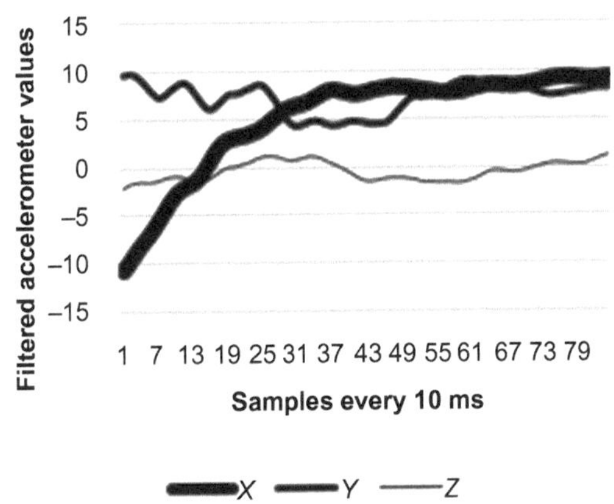

FIGURE 1.15

Dominancy change.

$$R_t = \sqrt{x^2 + y^2 + z^2} \tag{1.2}$$

where R is the resultant magnitude vector.

Fig. 1.16 shows the resulting magnitude. The unwanted vibrations are further smoothened by applying a first-order exponential weighted average filter as shown in Eq. (1.3), where s is the filtered magnitude and α is smoothing factor, which is chosen as 0.1 for best results.

The filter output modifies the input magnitude by a weighted filter coefficient (i.e., adding some weighted percentage [10%) of the magnitude). The resultant filtered data is denoted as s. Fig. 1.16 (waveform of filtered magnitude), shows the

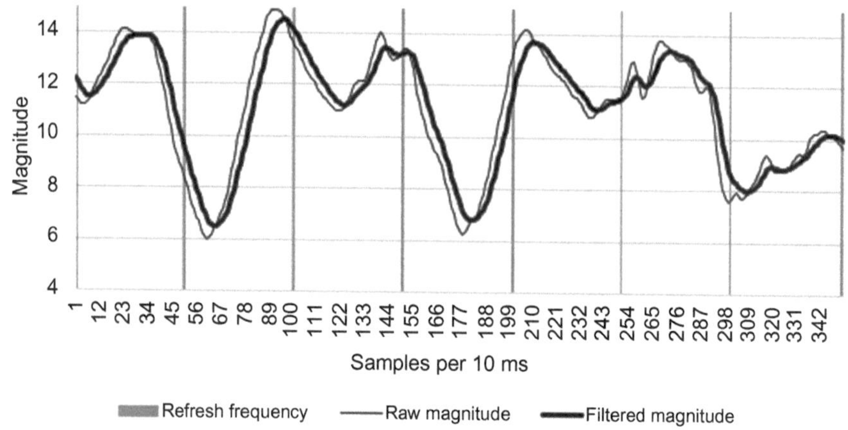

FIGURE 1.16

Magnitude filtration.

$$s(t) = \alpha R(t) + (1 - \alpha)s(t-1) \tag{1.3}$$

waveform with one maximum value and uneven minimum value. This waveform shows that a step is defined when a wave moves from either the negative to the positive peak or vice versa. In order to efficiently calculate the peak we need peaks for refresh period. Thus a refresh frequency of 500 ms is considered (i.e., 50 samples). Considering the data per refresh period shown in Figs. 1.17 and 1.18 the algorithm will calculate peak as follows:

$$v_{p-p} = s_{max} - s_{min} \tag{1.4}$$

If we consider every $V_{p\text{-}p}$ for further calculation of steps then we need to consider all activities of daily body vibrations. To calculate the minimum $V_{p\text{-}p}$ precision the major factors of activity of daily living and walking are considered. If the new result of $V_{p\text{-}p}$ is greater than the minimum change value (i.e., precision), then this $V_{p\text{-}p}$ is in the desirable region of walking. The choice of the right precision value varies from person to person. For dynamic computation of precision in the IWD algorithm, 10 levels of precision are considered between every $V_{p\text{-}p}$ (precision = $V_{p\text{-}p}/10$). All other small body vibrations are neglected at this point.

To compute the positive and negative peaks from $V_{p\text{-}p}$ a reference point is required (i.e., the threshold). Since $V_{p\text{-}p}$ is dynamic 50% of $V_{p\text{-}p}$ is considered as the ideal threshold: $T_h = V_{p\text{-}p}{*}0.5$. The two regions are considered as high threshold region as 75% of $V_{p\text{-}p}T_{high} = V_{p\text{-}p}{*}0.75$ and low threshold region as 25% of $V_{p\text{-}p}T_{low} = V_{p\text{-}p}{*}0.25$. Now the unit step is considered as the region (high/low threshold) of old and new magnitude (i.e., when the old magnitude was in the low region and the new is in the high region then it is a negative-to-positive peak transaction step and vice versa).

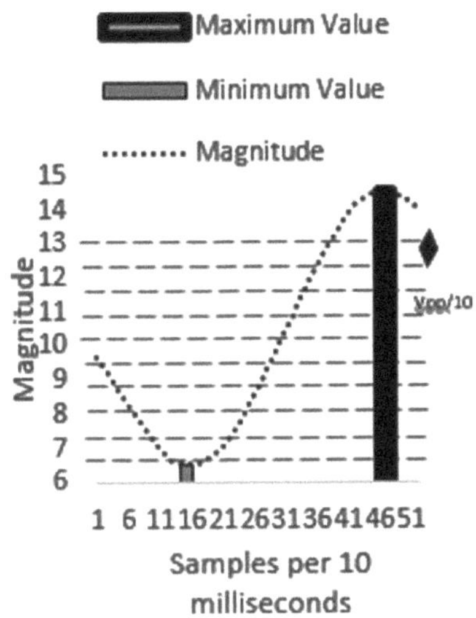

FIGURE 1.17

Magnitude refresh peak.

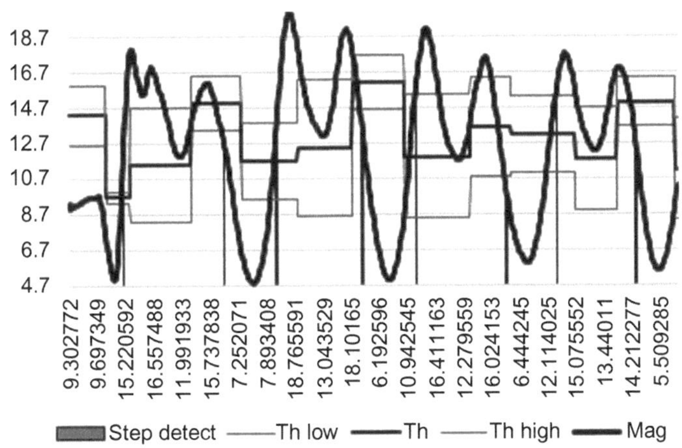

FIGURE 1.18

Dynamic threshold.

There is high possibility of detection of false step, if shaking of hands of human is computed rather than legs to compute step. When the pedometer vibrates rapidly or slowly due to any cause other than walking or running, the step counter will also take it as a step. Most false steps can be eliminated by considering time window and count regulation. Count regulation determines whether steps are part of a rhythmic pattern. Suppose the steps are in regulation after count four, then the result is refreshed and displayed continuously after the fifth valid step. The time window is used to discard the invalid vibrations, assuming that people can run as rapidly as five steps per second and walk as slowly as one step every 2 seconds. The interval is defined as (0.15–2.0 seconds).

Generally a 100 Hz speed consumes a lot of power and thus is not power efficient, so in order to make it efficient wake on motion is implemented. "Wake on motion" means the hardware monitors the acceleration data and detects any large change over a short period of time. The threshold levels can vary between 4 and 1020 mg in 4 mg increments.

Since execution of the pedometer algorithm thread depends on the configuration of wake on motion, this mode cannot be configured in the same thread. If it disables the pedometer thread, there is no other thread to wake up the pedometer thread. A callback which is called to monitor when to wake up is called in ISR context, which is available as new data everytime.

The time to enable and disable the pedometer thread depends on the following points:

1. Time difference of two valid steps is greater than 1 second in running when dynamic threshold is greater than magnitude 18.9.
2. Time difference of two valid steps is greater than 2 seconds in waking when dynamic threshold is less than magnitude 16.0.
3. Time difference of current time and last valid step time is greater than 2 seconds.
4. If valid steps are less than 5 in particular time window, that is the person is not in continuous motion.

Functions

```
enum cmd{
        DISABLE_MOTION=0,
        ENABLE_MOTION,
}
static void dataReadyHISR_cb(void)
{
        if (cmd==DISABLE_MOTION)
        {
                stopPedoThread();
                motion_on_lp(500,1,1);
        }
        else
        {
                /*Restore previous configuration*/
                motion_on_lp(0,0,0);
                /*wake up code for pedometer */
                startPedoThread();
        }
}
```

Code snippet1 ISR thread-Motion on detect on Invensence MPU9250 this is function which is called to detect the motion.

1.7 Conclusion

The following test cases were examined to verify the results and to make the device more power efficient. These are the changes observed while testing on a fully charged IWD smartwatch with a 320 maH battery.

Test case	Pedometer without wake on motion	Pedometer with wake on motion
Case I—screen OFF	12 h	48 h
Case II—all time screen, Bluetooth ON	6 h	24 h
Case III—only pedometer ON	24 h	More than 50 h

Case I is when the screen of the device is OFF; case II is when the screen and Bluetooth are ON all the time, and case III is where only the pedometer is ON (screen and Bluetooth OFF).

The following cases were also considered:

1. Woman walking in high heels as shown in Fig. 1.19; peaks between dynamic threshold high and threshold low are discarded.
2. Walking with hand in pocket as shown in Fig. 1.20; discard magnitude changes lower than precision value 4.
3. Region of activity is considered for acceleration while driving using pitch, with an interface to user to select hand.

Fig. 1.21A and B shows snapshots of the mobile app recording the step count using the Virtuino IoT platform. It also indicates whether the device is working in power-saving mode or not. The power overload mode "LED"

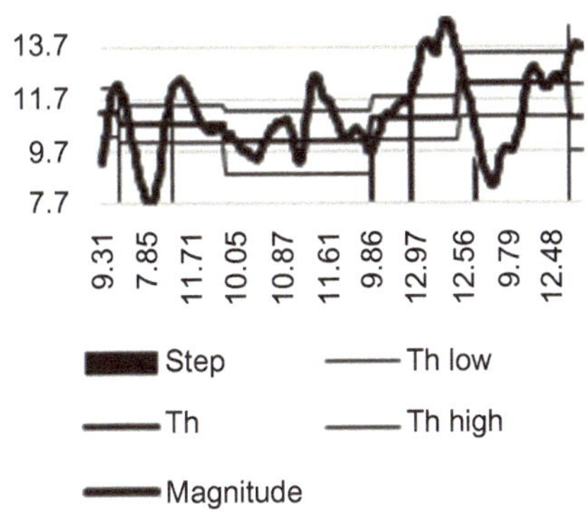

FIGURE 1.19

Woman walking in high heels.

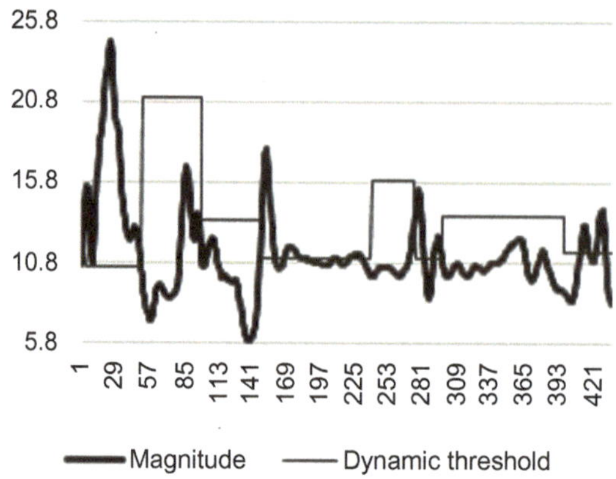

FIGURE 1.20

Walking with hand in pocket.

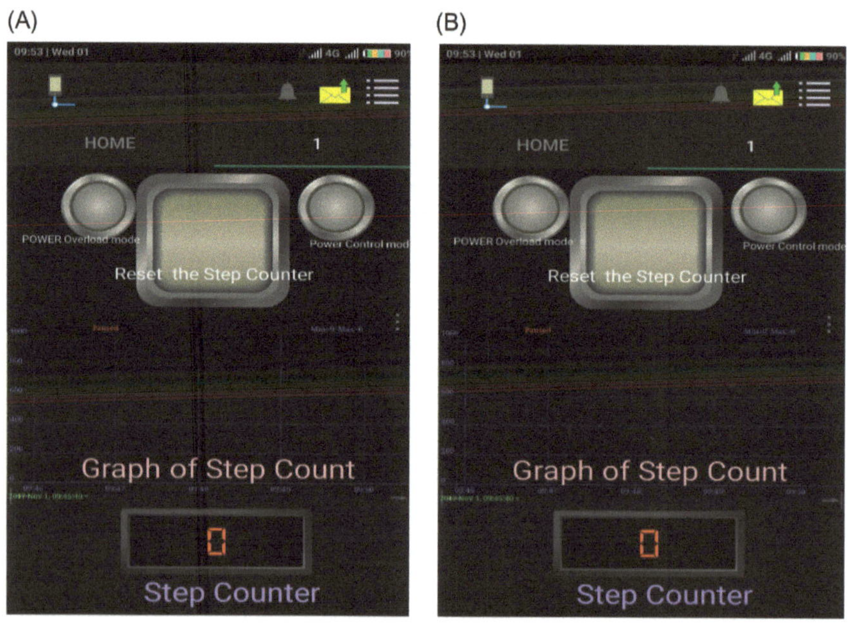

FIGURE 1.21

(A) Snapshot of developed app to record IWD data view1. (B) Snapshot of developed app to record IWD data view2.

indicates the IWD system is using more power than expected. The power control mode LED indicates power-saving mode is working and the system is using power *efficiently*. The graph of the step count shows the recorded pattern of data hourly, weekly, and monthly. The step count window shows the present step count status, which can be reset using reset the step counter button.

Fig. 1.22, Fig. 1.23, and Fig. 1.24 show the readings form the accelerometer of the *X*-axis, *Y*-axis, and *Z*-axis on the ThingSpeak cloud server, respectively. Fig. 1.25 shows the data capture from the step counts of the accelerometer.

The overall conclusion is that the IWD with the pedometer algorithm is the more power-efficient device for different cases.

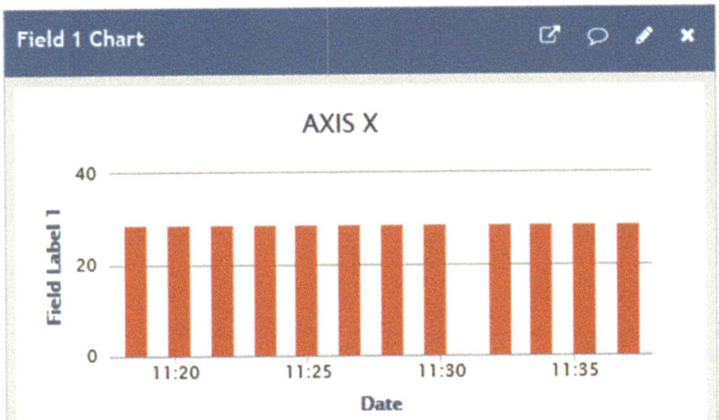

FIGURE 1.22

Data capture from accelerometer *X*-axis.

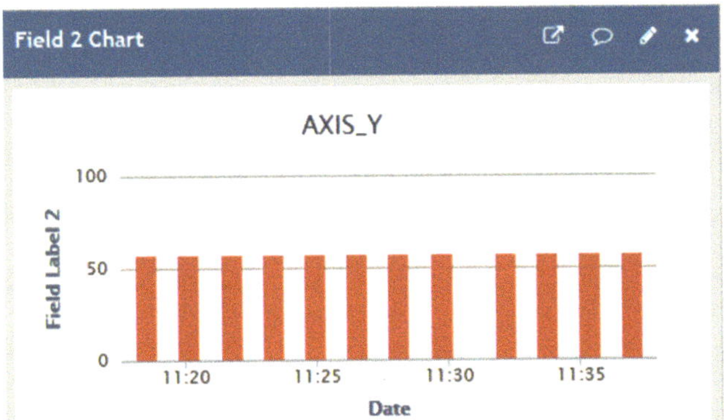

FIGURE 1.23

Data capture from accelerometer *Y*-axis.

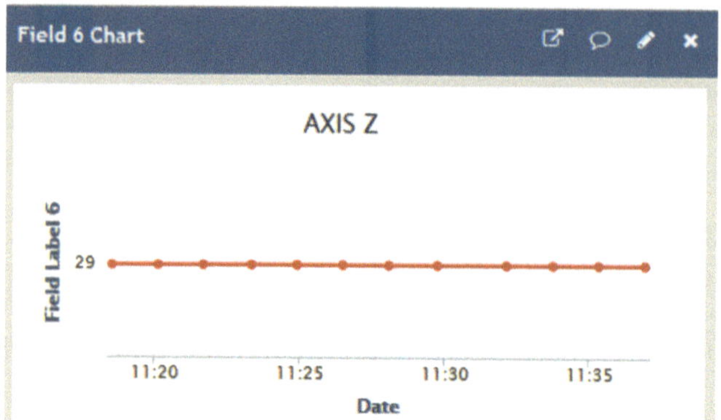

FIGURE 1.24

Data capture from accelerometer *Z*-axis.

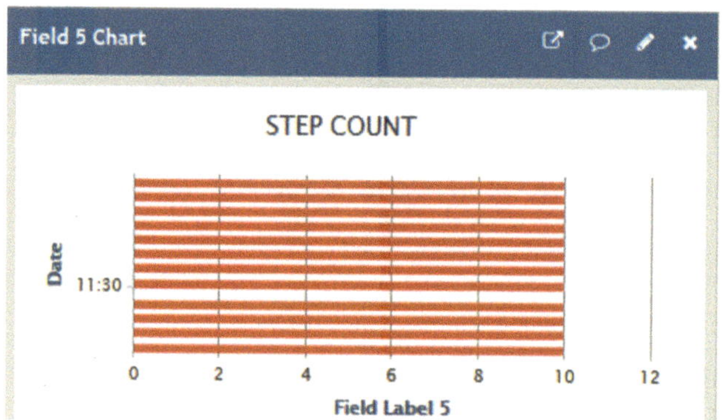

FIGURE 1.25

Data capture from step counts.

References

[1] L.F. McCarthy, 10,000 steps a day to a healthy heart, Better Nutr. 64 (2) (2002) 3–4.

[2] N. Zhao, Full-featured pedometer design realized with 3-axis digital accelerometer, Analog Dialogue 44 (6) (2010) 1–5.

[3] G.S. Kolt, G.M. Schofield, N. Kerse, N. Garrett, T. Ashton, A. Patel, Healthy steps trial: pedometer-based advice and physical activity for low-active older adults, Ann. Family Med. 10 (3) (2012) 206–212.

[4] B. Ewald, M. McEvoy, J. Attia, Step counts superior to physical activity scale for identifying health markers in older adults, Br. J. Sports Med. (2008).

[5] N. Abhayasinghe, I. Murray, A gyroscopic data based pedometer algorithm, in: 2013 8th International Conference on Computer Science & Education (ICCSE 2013), Colombo, Sri Lanka.

[6] Jayalath, N. Abhayasinghe, A gyroscopic data based pedometer algorithm, in: Computer Science & Education (ICCSE), 2013 8th International Conference on, IEEE, 2013, April, pp. 551–555.
[7] Walkmeter, Pedomter—walking, running, hiking, Abvio Inc., 2017 (Version 10.6.10) (Mobile application software). Retrieved from: <https://itunes.apple.com/us/app/walkmeter-gps-pedomter-walking-running-hiking/id330594424?mt = 8>.
[8] H.M. Husted, T.L. Llewellyn, The accuracy of pedometers in measuring walking steps on a treadmill in college students, Int. J. Exer. Sci. 10 (1) (2017) 146.
[9] Pedometer, Wikipedia, The Free Encyclopedia. Retrieved from: <https://en.wikipedia.org/w/index.php?title = Pedometer&oldid = 787545264>, 2017, June 26 (accessed 20.07.17).
[10] Tomlein, M., Bielik, P., Krátky, P., Mitrík, S., Barla, M., &Bieliková, M. (2012). Advanced pedometer for smartphone-based activity tracking, in: HEALTHINF, pp. 401-404.
[11] N. Jahan, S. Shenoy, Relation of pedometer steps count & self reported physical activity with health indices in middle aged adults, Diabet. Metabol. Syndr. Clin. Res. Rev. (2017). Available from: https://doi.org/10.1016/j.dsx.2017.07.033.
[12] M. De Craemer, et al., Validity of the Omron pedometer and the actigraph step count function in preschoolers, J. Sci. Med. Sport 18 (3) (2015) 289–293.
[13] R.D. Burns, et al., Establishing school day pedometer step count cut-points using ROC curves in low-income children, Prevent. Med. 86 (2016) 117–122.
[14] Y. Feng, et al., Comparison of tri-axial accelerometers step-count accuracy in slow walking conditions, Gait Posture 53 (2017) 11–16.
[15] A.K. Singh, et al., Accuracy of the FitBit at walking speeds and cadences relevant to clinical rehabilitation populations, Disability Health J. 9 (2) (2016) 320–323.
[16] Acceleromter Wikipedia, The Free Encyclopedia. Retrieved from: <https://en.wikipedia.org/w/index.php?title = Acceleromter &oldid = 787517520> 2017, June 25 (accessed 20.7.17).
[17] M. Pedley, Tilt sensing using a three-axis accelerometer, Freescale Semiconductor Application Note 1 (2013) 2012–2013.
[18] L.P. Toth, et al., Video-recorded validation of wearable step counters under free-living conditions, Med. Sci. Sports Exer. 50 (6) (2018) 1313–1314.
[19] A.A. Alhusaini, et al., Pedometer-determined physical activity levels of healthy children and children with Down's syndrome, Somatosens.Motor Res. 61 (2018) 1–7.
[20] C.M. Coelho, et al., Effects of an unsupervised pedometer-based physical activity program on daily steps of adults with moderate to severe asthma: a randomized controlled trial, J. Sports Sci. 36 (10) (2018) 1186–1193.
[21] H.T. Binh, N.T.H. Thanh, N. Dey, Improved cuckoo search and chaotic flower pollination optimization algorithm for maximizing area coverage in wireless sensor networks, Neural Comput. Appl. (2016) 1–13.
[22] S. Roy, et al., Symmetric key encryption technique: a cellular automata based approach in wireless sensor networks, Proc. Comput. Sci. 78 (2016) 408–414.
[23] C. Bhatt, N. Dey, A.S. Ashour (Eds.), Internet of Things and Big Data technologies for next Generation Healthcare, Springer, Cham, 2017.

Further reading

Exponential smoothing, Wikipedia, The Free Encyclopedia, 2017, July 5, Retrieved July 20, 2017.

CHAPTER

2 Biosensors and Internet of Things in smart healthcare applications: challenges and opportunities

Maria Pateraki[1], Konstantinos Fysarakis[2], Vangelis Sakkalis[1], Georgios Spanoudakis[2], Iraklis Varlamis[1,3], Michail Maniadakis[1], Manolis Lourakis[1], Sotiris Ioannidis[1], Nicholas Cummins[4], Björn Schuller[4], Evangelos Loutsetis[1] and Dimitrios Koutsouris[5]

[1]*Institute of Computer Science, Foundation for Research and Technology–Hellas, Greece*

[2]*Sphynx Technology Solutions AG, Zug, Switzerland*

[3]*Department of Informatics and Telematics, Harokopio University of Athens, Athens, Greece*

[4]*ZD.B Chair of Embedded Intelligence for Health Care and Wellness, University of Augsburg, Germany*

[5]*Institute of Communication and Computer Systems, Athens, Greece*

Chapter Outline

Wearable and Implantable Medical Devices. DOI: https://doi.org/10.1016/B978-0-12-815369-7.00002-1

2.1 Introduction

As the global population ages, several disorders, diseases, and impairments are becoming prevalent with significant financial and social implications. The prevention and appropriate handling of such conditions will benefit both individuals and national healthcare systems. Smart information and communications technology (ICT) solutions can provide benefits such as improving the quality of life and supporting independent living of the elderly and other groups with chronic or acute conditions.

Biosensors are complete analytical devices composed of a receptor, an active biological factor that can be an enzyme, antibody or similar, coupled with a transducer and are able to detect the existence of a particular analyte. They are continuous sources of biological data and when integrated in a smart analytics framework, they can provide useful information about a person's condition. Smart biosensors in health capitalize on recent advances in microtechnology and wireless communication in order to collect and transmit information and, in conjunction with actuators, are able to provide better monitoring and treatment. For example, wearable biosensors provide vital signs monitoring for everyone, including patients, children, and the elderly, and are very effective in health risk prevention and control.

The use of smart biosensors can benefit people that require medical support or care, as well as for older adults, who face a gradual degradation of their motor, cognitive, and other skills due to aging. Smart biosensors can also help individuals exposed to harsh working environments, or who perform stressful tasks or tasks that induce health risks. The continuous monitoring of biosignals in combination with environmental sensing allows smart systems to promote a reactive living and working environment that provides appropriate and timely recommendations, acts preventively, and mitigates health risks.

The long list of biosensor applications [1,2] includes, among others, helmets for treating depression through electrical pulses and smart clothes (e.g., socks, shoes, t-shirts, and smart vests) that noninvasively collect and transmit vital signs and integrate well with the concept of a smart house or a smart working environment.

Several technological and societal challenges pose barriers to the widespread adoption and deployment of smart biosensor applications. These barriers relate primarily to the interoperability and expandability of the overall solution. Moreover, the ability to collect and process large amounts of data from heterogeneous sensors, as well as the ability to provide high-level data analytics and extract knowledge, ought to overcome all technological barriers. This ability to collect and analyze data will support the development of solutions and provide evidence supporting the effectiveness of personalized interventions and recommendations that will promote user trust. This chapter aims to highlight challenges and opportunities pertaining to the application of smart biosensors in healthcare and to present the main building blocks required for such applications. Finally, three state of the art solutions leveraging smart sensors in this context are presented.

The chapter is further structured as follows. Section 2.2 highlights health challenges faced by the elderly, older workers, and infants, as well as the social and financial implications associated with these challenges. Section 2.3 presents the main challenges and opportunities for technology-enabled care (TEC). Section 2.4 describes the building blocks for the Internet of Things (IoT) and the Internet of Medical Things (IoMT) in health and well-being applications, which are smart environment enablers, back end enablers for personalized recommendations, plus security and privacy enablers. Section 2.5 presents three state of the art smart healthcare applications to address challenges in living and working environments.

2.2 Health challenges for the elderly, older workers, and infants

According to HelpAge International,[1] by the year 2050, one in five people will be over 60. Among the 10 most prevalent health challenges that lead elderly people to physical injuries and affect their ability to have a healthy and independent life [3] are the following:

- *Hearing loss (HL)*: HL is one of the most prevalent chronic conditions in older adults [4] and affects one-third of people ages 65–74, as well as nearly half of those older than 75. HL increases the risk of cognitive decline and depression and can lead to social isolation.
- *Cardiovascular diseases (CVDs)*: Hypertension, ischemic heart disease, and heart failure are the primary causes of death globally (31% in 2016, 85% of which were from ischemic heart disease and stroke) [5]. CVDs affect many aspects of life for elderly adults including their physical, social, and emotional status.
- *Cognitive impairments (CIs)*: CIs affect the ability of people to think, learn, and remember. Dementia is the most common issue with approximately 47.5 million cases worldwide and a prediction to triple by 2050. Dementia has high comorbidity with heart failure among the elderly and affects several cognitive domains including executive and motor function [6].
- *Mental health (MH) issues*: MH issues affect a significant proportion of the older population (e.g., depression at 7%, anxiety disorders at 3.8%, substance

[1] www.helpage.org/resources/ageing-data/global-ageing-statistics/

use problems at nearly 1%, self-harm attempts underlie 25% of elderly deaths) and in some cases have an impact on physical health. Adults with other health conditions (e.g., heart disease) have a higher risk for MH issues.

- *Balance disorders (BDs)*: BDs are disturbances in coordination that make someone feel unsteady, dizzy, or have a sensation of movement. This progressive, age-related loss of sensory functions and inability to control body movements frequently lead to falls, physical injury, and death (one elderly person dies from falling every 29 minutes [7]). Frailty is quite prevalent among elderly people. The prevalence ranges from 33% to 88% depending on how frailty is defined and steadily increases with age. Frailty and other progressive disorders increase the risk for the elderly and negatively affect their quality of life [8].

Because health problems increase with age and the age limits for retirement continue to increase, more employees are likely to develop health problems while still at work. Changes in physical abilities (e.g., balance, mobility, dexterity, stature, strength, and aerobic power) can result in a reduced tolerance of physical work and declines in motor skills (e.g., difficulty maintaining coordination, loss of flexibility), rendering tasks that require fine manipulation harder for older workers. Finally, changes in cognitive abilities (e.g., declines in episodic memory, executive functioning, and attentional control) reduce task performance and affect work capacity [9]. Conversely, cumulative exposure to demanding work can have a significant impact on health and functional abilities, wellness at work, and productivity. Arduous working conditions (e.g., exposure to extreme temperatures, dangerous substances, or noise) have a major influence on the risk of developing work-related ill-health (e.g., illness, stress, fatigue). When work environments require workers to exert intense physical effort, even occasionally, it is important to keep workers safe and healthy so as to increase their resilience and avoid injury risks [10].

On the opposite end of the age scale, various pathological conditions can arise during the first year of life in infants, which call for immediate detection and intervention. Parents cannot always identify the signs of pathology that specific movements and sounds of an infant indicate. According to the World Health Organization[2] a child is at the highest risk of dying in the first 28 days of life, while at home. This risk is especially high when neonates are discharged early from the hospital. Current countermeasures include postnatal care plans built around home visits by healthcare professionals. Many pathological situations can occur during an infant's sleep that can potentially be harmful to the infant's health if not detected promptly: breathing disorders, vomiting, arrhythmias, epileptic and febrile convulsions, high fever, and sleep disorders including sudden infant death syndrome (ranked in order of life-threatening risk).

[2] https://www.who.int/news-room/fact-sheets/detail/newborns-reducing-mortality

2.3 Challenges and opportunities for technology-enabled care

2.3.1 Low-cost technology

The explosive growth of IoT [11] and associated technologies has placed downward pressure on the cost of associated hardware and software [2]. Building successful low-cost TEC solutions depends on a number of factors related to sensors or the network itself. Sensors must have: (1) low power consumption for prolonging battery life; (2) physical characteristics that provide unobtrusiveness; (3) robustness to minimize maintenance; (4) wireless connectivity to facilitate networking using widely accepted standards; and (5) data preprocessing capabilities to reduce computational load on gateways and the cloud. The network itself must support: (1) the deployment and management of a large number of low-cost sensors and (2) high processing speed and portability to enable better care at lower cost.

2.3.2 Modular, interoperable, expandable solutions

The integration of a large number of heterogeneous smart objects that use different communication technologies (e.g., Bluetooth, RFID, Zigbee, 802.11, 802.15.4), run a variety of often proprietary protocols and applications, and have limited exposed interfaces, requires careful engineering actions that will interconnect IoT devices in a network and will build realistic and useful novel IoT solutions [12]. Many surveys highlight that vendor lock-in and complicated security and management processes hinder the broader adoption of IoT technologies [13]. Aiming to alleviate the interoperability issues, various emerging IoT platforms are either domain-specific (e.g., UniversAAL[3]) or general purpose (e.g., FI-WARE,[4] GoogleApp engine[5]). Thus, developers rely on existing platforms and their services, and future platforms must be interoperable with the existing ones. Moreover, various standardized "IoT communication protocols" have been proposed, aiming to address the interoperability and fragmentation issues. The MQ Telemetry Transport (MQTT[6]) is one such machine-to-machine (M2M) connectivity protocol, which was recently standardized by OASIS (also standardized as ISO/IEC 20922), and already applied in various domains including eHealth [14] and smart homes [15]. The OASIS standard devices profile for web services (DPWS) [16] supports the interaction with resource-constrained devices and has been studied extensively in many areas, including eHealth and smart homes [17]. The IETF standard constrained application protocol (CoAP) [18] offers an

[3] http://www.universaal.info
[4] http://www.fi-ware.org
[5] https://cloud.google.com/appengine/
[6] https://docs.oasis-open.org/mqtt/mqtt/v5.0/mqtt-v5.0.html

alternative web transfer protocol that allows the integration of constrained IoT nodes through lightweight interactions.

2.3.3 Big data and machine learning

With environmental sensors and wearable devices, smart healthcare platforms can continuously monitor the health status and activities of patients and the elderly, as well as the safety and security of the environment [19,20]. This continuous flow of data in combination with individual's medical histories, can support personalized diagnosis and assistance and can automate important tasks, such as medical data archiving and evaluation of the effectiveness of medical interventions. Recent advances in artificial intelligence (AI) and big data processing technologies have allowed the implementation of highly reliable, accurate, and robust infrastructures for data recording and processing [21].

The analysis of data from wearable devices is led by decision-making process requirements, which define the data to be collected and extracted in support of intelligent decision-making. To support intelligent decision-making, the following requirements must be considered: (1) modular and interoperable data ingestion in which devices from different manufacturers contribute to a common data model, (2) parallelization and data stream processing in all steps from data acquisition to storage and processing, (3) use of data analytics and AI to support automatic monitoring and decision-making, (4) privacy-aware data processing to ensure consistent data encryption, database security, as well as secured communication channels.

2.3.4 Security and privacy

Two main issues regarding security and privacy of relevant applications are the complex interaction schemes that take place during typical, everyday use of smart healthcare solutions (e.g., patients/users with their caregivers/medical professionals), along with the private/sensitive nature of the handled data. These factors necessitate the integration of strong security and privacy provisions, including seamless authentication and authorization services, for the protection of the framework's M2M and machine-to-human (M2H) interactions [21]. IoT devices have been designed mainly considering low-cost, low-energy usage, ease of setup and use, and interconnection, but not security. Since health monitoring systems or even smart homes and workplaces may include sensitive assets, it is important to protect them from malicious attackers.

The below define a set of security and privacy risks and considerations:

- Small, low-cost, interconnected devices have immature security functions.
- Low processing capabilities of the network require computationally intensive real-time tasks (e.g., condition reasoning) to be moved to the cloud.

- Secure communication within the smart home uses a range of protocols (WiFi, Bluetooth, NFC, ZigBee, and others) that may open various exploits.
- Privacy protection systems (e.g., smart home sensors) generate a large amount of highly personal data and metadata.
- Consent for secure sharing of anonymized information is necessary.
- The physical security of a smart home is linked to the safety of sensitive systems for the occupants' healthcare.
- Communications with the back end processing systems must be secure.
- Secure and privacy-preserving interactions (communication, processing, storage) with third parties (hospitals, clinics, etc.), their data and metadata.
- Authentication, authorization, and accounting mechanisms must be tailored to all actors (from elderly and clinical experts to sensors and back end systems).
- Reliability and availability of information from back end databases and real-time data streams must be guaranteed.

The information sources to be protected encompass data from smart devices, including raw data, logs, metadata (headers, content type, dates, etc.), events (alerts, warnings, errors, etc.), rules, settings and preferences (which may disclose information about the end user's conditions), updates to and from smart devices, postprocessed data, and interactions of the smart home with the back end cloud, as well as interactions of the back end cloud with various healthcare service providers. All adopted security and privacy mechanisms must be tailored to the above requirements. A key characteristic of these mechanisms will be their capability to adapt in real time to a variety of usage requirements (e.g., context, privacy preferences, risk profile, and other parameters).

2.4 Internet of Things and Internet of Medical Things building blocks for health and well-being applications

2.4.1 Smart environment enablers

2.4.1.1 Wearable and assistive medical devices

A basic component of a smart solution for health is a wireless sensor network that obtains automated, continuous, and real-time measurements of physiological signals and performs limited data processing and functions. Vital signs, such as heart rate, heart rate variability, body temperature, skin conductance, respiration rate, blood pressure, blood glucose, oxygen saturation, as well as activity related signals can be captured and analyzed using appropriately selected sensors that can be placed over clothes or directly on the body. Another physiological measurements that can be recorded is the hearing response, which is supported by hearing aids. Current state of the art technology in the aforementioned sensors lacks multiparameter systems for the concurrent monitoring of multiple physiological

Table 2.1 Physiological measurement solutions in the Internet of Things era.

Physiological measurement	Sensor type	Data provided
Heart rate	PPG	Raw
Blood pressure	Pressure sensor	Raw
Blood glucose	POC	Raw
Respiration rate	Sensitive stretch sensor	Raw
Oxygen saturation	PPG	Raw
Body temperature	Thermocouple	Raw
Skin conductance and temperature	GSR	Raw
Activity	IMU	Raw, aggregated
Sympathetic nervous system activity	GSR	Raw
Hearing response	–	Raw, aggregated

GSR, *Galvanic skin response;* PPG, *photoplethysmography;* POC, *point-of-care;* IMU, *Inertial measurement unit.*

measurements. On the other hand, the utilization of separate sensors is not practical and may cause inconvenience and obtrusive operation for the end users.

Physiological signals from the human body can be measured with various sensor technologies [22], some of which are depicted in Table 2.1. The heart rate, which has become a routine measurement, can be easily extracted from photoplethysmography (PPG) signals. Blood pressure derives from inflatable cuffs accompanied with a stethoscope. The evolution of this medical device resulted in an integrated smart pressure sensor [23]. Moreover, blood oxygen saturation, a valuable vital parameter, can now be easily measured through the exploitation of PPG technology. PPG is a biophotonic technology using two different light wavelengths [24]. The basic type of sensors in skin sweat monitoring is epidermal galvanic skin response (GSR) sensor. Respiration rate is used to detect stress and potential hypoxia [25].

More recently, wearable devices that can infer the human physical activity have gained popularity. Such devices often incorporate inertial measurement unit (IMU), global positioning systems (GPS), PPG sensors, ECG leads, and sophisticated firmware capable of high quality and continuous biosignal monitoring [26]. Finally, sympathetic nervous system activity can be captured using electrodermal activity sensors, which offer information about alterations in the central nervous system. Assessment of these alterations leads to indicators of the emotional condition of the subject [27]. The abovementioned sensors can be integrated into wearable devices specifically designed to extract raw, aggregated, or both types of data and collectively compose a smart IoT ecosystem.

Considering the above landscape, a smart healthcare solution must be able to integrate physiological measurements, such as those reported above, aggregate (and preprocess, if necessary), and transmit them to a back end cloud platform. There, huge amounts of raw and aggregated data can be analyzed through advanced big data analytics in order to understand behavioral activity, detect probable risks, and provide adequate interventions.

2.4.1.2 Mobile devices

Mobile devices are an integral part of smart healthcare solutions. They are the near perfect interface to actively gather self-reported data from individuals. Additionally, they contain a vast array of embedded sensors and features for collecting a large variety of data, both actively and passively, which can be used to infer information regarding a subject's current health or mental state. Behavioral signals such as speech, facial expression, and gaze can also be collected through the cameras and microphones embedded in all consumer smartphones and tablets. Furthermore, mobile devices offer temporary storage and the means to remotely transmit this information. They also represent a straightforward solution for the intermediate storage of health and wellness data collected from wearable and other IoT devices before transmission to a back end cloud platform for analysis [28]. However, as already pointed out in the previous section, this transmission represents a potential security risk and can quickly drain the limited power available to these devices. In addition to functioning as storage, transmission, and potentially processing devices, smartphones represent a new source of health and wellness information. In particular, the shift to mobile devices, smartphones, and tablets as core communication platforms, has resulted in a new source of data known as digital-trace information. This data stream is generated implicitly through smartphone usage and can be collected passively and unobtrusively (without specific user interaction) by the use of specially designed apps. One such app is RADAR-BASE, which runs as a background process and automatically collects and transmits this information for analysis and predictive monitoring [29]. Implicit trace information gathered from smartphones includes social activities as monitored via call and message logs, social media usage, or Bluetooth connectivity, and activity levels as inferred from embedded sensors or GPS data. Ambient noise and light levels, screen time, and application usage can also be easily collected.

A growing area of research in smart healthcare solutions is the embedding of AI technologies directly into mobile devices. However, considering that a modern deep neural network can have millions of hyperparameters to tune, the computational demands associated with these technologies are very high, potentially requiring hundreds of megabytes. In several cases they also need substantial data movement to support their operation, thus constituing a highly nontrivial process. One growing research direction within neural networks is the development of approaches that can import large networks and optimize them until they are executable on a low resource smart device [30]. Other methods are aimed at lowering the memory footprint and computational complexity of AI technologies while maintaining reasonable accuracy. Developing low resource networks increases the likelihood of smart systems being able to run offline, increasing user privacy and reducing energy consumption concerns associated with transmission bandwidth, all of which are core considerations for a robust smart healthcare solution.

2.4.1.3 Environmental monitoring and Internet of Things platforms

Environmental monitoring encompasses a broad variety of IoT applications that involve online monitoring of environmental parameters such as temperature, humidity, noise levels, air pollutant concentrations, etc., which affect people's safety and well-being [31]. The measured parameters are collected through dedicated gateways by an IoT platform for monitoring and analytics.

The most popular IoT platforms for use as secure gateways are AGILE IoT, Eclipse Kura, and HomeAssistant.[7] They are open source, feature ready-to-use field protocols, and support wireless and wired IoT networking technologies such as WiFi, Bluetooth Low Energy (BLE), ZigBee, Z-Wave. Technology advancements in Bluetooth low energy, make BLE devices suitable for the development of IoT networks, combined with power harvesting elements and mobile gateways [32]. For publishing data and events to IoT cloud platforms, MQTT connectivity is an option available to all platforms. Every platform uses its own authentication system (OAuth2, multi factor authentication, etc.). For secure access, HTTP SSL/TLS protocols and MQTT connectivity ensure the privacy of established connections.

AGILE IoT builds a modular and adaptive gateway for IoT devices that supports interoperability of devices and data. Modular hardware solutions that adopt all communication protocols in combination with the appropriate software components that offer smart services (data management on the gateway, intuitive interface for device management, etc.) allow fast prototyping of extensible solutions. Eclipse Kura is an extensible open source IoT Edge Framework that offers application programming interface (API) access to the IoT gateways (I2C, GPS, GPIOs, serial ports, etc.). HomeAssistant is an open source IoT platform with hundreds of built-in components for connectivity with off-the-shelf sensors, providing an easy framework for importing more devices and a mobile-friendly interface for setting up automation rules and monitoring devices.

2.4.1.4 Camera-based monitoring of humans

Despite not yet being widely adopted in IoT frameworks, visual sensing using cameras has several attractive advantages over other sensing modalities. These advantages stem from the fact that visual sensing can support the extraction of detailed context information from a scene, while being passive, low cost, and nonintrusive. Context awareness facilitates a better understanding of the activities/actions, health, and risks faced by a subject being monitored by detecting behavior patterns and supporting more precise inferences about the subject's situation and environment. Many systems rely by design on the extraction of low-level context information, such as the location of users, derived by nonvisual sensors and technologies. However, in cases with more elaborate monitoring

[7] http://agile-iot.eu, https://www.eclipse.org/kura and https://www.home-assistant.io

requirements, for example, when one needs to extract higher level information such as behavioral patterns and the subject's activity, or when the environment is occupied by multiple persons or contains certain materials such as metal parts that may interfere with localization radio signals, visual information from camera sensors can provide richer and more precise information. However, in an IoT camera-based monitoring system, there are security and privacy risks that relate to the transmission of images away from the imaging sensor for processing. Therefore it is preferable to move the application of security and privacy protection closer to the sensor, enhancing control of data privacy and simultaneously accommodating key concerns among users regarding privacy violations.

Beyond privacy issues, considering that human behavior in daily activities is complex and highly diverse, monitoring such activities presents significant challenges. As outlined in [33] these challenges are: (1) recognizing concurrent activities (i.e., individuals performing several activities simultaneously), (2) recognizing interleaved activities (i.e., activities that are overlapped with others), (3) ambiguity of interpretation (i.e., similar actions may be interpreted differently depending on the context), and (4) support of multiple users (i.e., recognize activities performed in parallel by many users in a group). Human behavior is characterized by varying time frames and levels of semantics [34]. In addition to the above, robustness to variations in real-world indoor and outdoor environments is affected by scene- and image-dependent factors, such as variations in the performance of actions, background clutter, occlusions, lighting conditions, and camera sensor selection and placement [35].

With the advent of low-cost, real-time dense depth cameras such as the Kinect,[8] numerous important approaches to action recognition and tracking problems have emerged, pushing the state of the art significantly forward [36]. Nevertheless, and despite the fairly accurate performance of state of the art algorithms in controlled or semicontrolled settings, coping with complex, realistic scenarios exposes the limits of these algorithms, particularly effective handling of longer duration occlusions, which remains an unsolved problem in most current approaches [37]. Lastly, such approaches suffer from natural light interference and limited range, and hence are restricted to indoor environments. On the other hand, passive stereo cameras have a wider range of application, as they can operate in sunlight and their field of view can be adjusted by using different cameras, lenses, or baselines.

Apart from spatial ambiguities related to human body segmentation in complex scenes, ambiguities in the temporal domain may also affect action recognition. These are easily resolved with repetitive actions, but they may greatly affect the detection of nonrepetitive actions such as pulling, pushing, or lifting an object. Moreover, performance may degrade in case of domain shift problems, for instance when the scale and shape of the human action are inconsistent with those of training data. Empirical results suggest [38] that convolutional neural network

[8] https://www.xbox.com/en-US/kinect

(CNN)-based algorithms are able to learn similar features between different actors performing the same action (i.e., performance nuance). However, in many real-world problems (e.g., surveillance scenarios), it is not possible to provide massive amounts of training data nor avail enough time for training. Thus, there is a need for algorithms that can work reliably in real time with moderate amounts of data and progressively improve their confidence as more data is learned, ideally in an unsupervised fashion.

2.4.2 Back end enablers for personalized recommendations

2.4.2.1 Knowledge abstraction for user profiling and temporal reasoning

In order to reduce predictable acute health episodes, a system should focus on eliminating complications, preventive disease management, and timely detection of anomalies based on past events. However, a characteristic of ordinary computerized healthcare systems is the limitation of user participation in the decisions of the system. User-centered design has been recently adopted as a methodological tool to inform the development of modern health technology systems. Capitalizing on the use of IoT technologies and analytics, modern systems are able to infer hidden patient information and their own risk-related parameters. Constant monitoring of incoming data can be used to trigger warnings based on the identification or prediction of user-independent abnormal parameter values or the identification of crucial deviations from a patient's data profile, which may indicate the increase of a risk. Moreover, exploitation of past data is important for the delivery of personalized treatment based on predictive modeling techniques that will determine the expected treatment response for a certain patient. Still, the comparison between past and current data, which are frequently stored in the form of time series, is not straightforward. Accordingly, it is often necessary to develop abstracted pictures of current and past events, which are contrasted to reveal abnormalities. Dimensionality reduction is commonly used as an approach to develop simplified representations of the different cases (data sequences) and similarity-based comparisons between them to support time series retrieval and decision-making [39]. In recent years, temporal abstraction (TA) has been used as a method to derive high-level concepts from time stamped data [40]. The idea behind TA is to move from a point-based to an interval-based representation of data, which effectively summarizes the data into meaningful parts that are interpretable by the users of the system [41]. The evidence arising from the comparison of different cases is fed into decision models to identify and suggest interventions that either prevent the occurrence of risks or reduce their effect on patient health.

The use of big data analytics and the ease of aggregating and synthesizing anonymous patient clinical records facilitates the creation of custom cohorts and metrics to extract knowledge that can be transferred and applied across different patients and can be a valuable service to third parties [42]. Interestingly, besides building accurate models of disease progression and providing personalized medicine in clinical practice, big data analytics facilitates the integration of medical data with wearable devices and IoT smart sensors. These devices provide information on supplementary behavioral determinants of health and may crucially support the analysis of potential public health policies regarding such interventions at the regional, national, and international levels [43].

2.4.2.2 Context-aware recommendations

The most interactive part of a smart medical care solution based on biosensors emerges when the system recommends actions to the end user based on the information collected by the sensor ecosystem. When developing a recommendation system for a specific purpose, such as the improvement of physical or mental status, it is important to consider what actions to recommend and at which moment they should be addressed to the user. This defines the concept of context-aware recommendation systems (CARS), which take into account the user spatiotemporal environment, as well as other conditions such as the user status (standing, walking, driving) or physical (tired or energetic) and psychological (happy or sad) conditions. Sensors can be used to detect user context [44] and are the backbone of CARS that support health and medical care. For example, Casino et al. [45] propose a CARS that takes into account the health information of citizens and their preferences, combines them with the real-time information about weather and air conditions collected from smart city sensors, and recommends personalized path alternatives that fit each end user profile. The "Motivate" CAR system [46] used several recommendations (e.g., take a break from work, stretch, walk, cycle to a park, go to a museum) that promote social, physical, and mental balance and considered various context parameters including location, user agenda, weather, user profile, and time. "Let's exercise" [47] is another CARS that recommends physical activities. Additional approaches for motivating older people to engage in social and physical activities are presented in Ref. [48], which also proposes a CARS for suggesting social and other events that match user profiles. Biosensors can take CARS to the next level by introducing an additional context, the psychological. The detection of stress and arousal can improve the recommendation timing and increase their acceptance rate.

2.4.3 Security and privacy enablers

Basic security tasks such as mutual authentication, encryption, and data integrity remain challenging in IoT. Encryption using elliptic curves and signatures has been shown to be possible on embedded devices but may not be possible on every sensor or actuator [49]. Confidentiality and integrity protection mechanisms also require strong authentication and authorization mechanisms. This requires assigning an identity to sensors and actuators (i.e., a sensor must store some secret to authenticate to a field device). In the past this was, for example, solved with a second channel and user involvement [50] or using certificates [51]. However, all these solutions lack scalability and support for dynamic, unobtrusive smart environments. Concerning security and privacy at the back end, because smart healthcare applications require distribution and processing of sensitive data, they will need to adopt new distributed and/or collaborative paradigms of cloud computing. The obfuscation and anonymization of uploaded data [52] is a simple technique to prevent sensitive information leakage; however, this technique affects the data and makes it unusable for other applications. Fully homomorphic encryption [53], privacy-preserving encryption [54], and attribute-based encryption have been proposed for encrypting sensitive user data without limiting the functionality of cloud applications. However, cryptography alone cannot sufficiently preserve user privacy and thus other forms of privacy enforcement must be employed [55], such as proper identity and authorization management by specifying and enforcing security, access control, and privacy policies. Indeed, an ENISA report (ENISA, 10) on security and resilience of e-health infrastructures and services identifies access control as a very significant priority in securing applications. Among the studied authorization schemes proposed for systems with different requirements and properties, a cross-platform solution that meets the requirements of all types of embedded systems and provides interoperability is the eXtensible Access Control Markup Language (XACML) [56]), the de facto standard for specifying and evaluating access control policies [57]. Also supporting XACML extension are its privacy-aware features [58].

Another important aspect related to the above and which raises significant concerns is the interplay between machine learning techniques and privacy. More specifically, there is a recent trend to design machine learning models that are trained from IoT data, which raises many privacy concerns and ethical issues. When the data used to train the models is comprised of unfiltered data from the real world, there is the risk of learning the respective behaviors that exist in the data, which may result in strange or unethical behaviors. The research on security and privacy of big data analytics (BDA) models still devotes less attention to the impact of similar solutions that assume distributed architectures and BDA models for IoT [59]. Several researchers agree that the best trade-off between utility and disclosure risk can be found at the time of model inference when there exist real data to evaluate the data utility and the impact of its disclosure, as opposed to estimating the risk a priori [60]. BDA pipeline modules are owned and managed by multiple operators, each with its own interests and agenda; therefore, we cannot always postpone all disclosure control to the time of analytics computation. In this context, noninteractive randomization at the time of data acquisition, while decreasing utility, can provide maximum flexibility and best accommodate provisions for compliance with regulations, ethics, and cultural factors.

Considering the above, to address the security and privacy concerns, a state of the art IoT/IoMT healthcare solution must combine novel and standardized technologies to provide lightweight and usable mechanisms for the authentication of its entities (devices, applications, users, etc.) [61] and the protection of their resources through strong, unambiguous, and fine-grained authorization services. The XACML authorization engine can form the basis of this endeavor, developing dynamic authorization services and providing the necessary variables (operational or situational context, as well as privacy requirements and other scenario/use case peculiarities). Privacy-aware features can be embedded into the policy definitions. Developed solutions for back end security must allow the creation of secure and privacy-preserving communications within and from the cloud infrastructure to the smart home and healthcare service providers in an end-to-end manner. The privacy controls implemented can also include differential privacy and selective data obfuscation and randomization, both for raw data and for outcomes of the data analytics, learning, and evolution processes. The combination of the above guarantees visibility of the system's status and consequent enhanced operator control and accountability. The platform must provide a significantly higher level of security and privacy than what is currently available in the domain to unambiguously alleviate the pertinent concerns.

2.5 Smart healthcare applications—state-of-the-art research efforts

Within the landscape sketched in the previous sections and motivated by the significant benefits of IoT/IoMT-enabled smart healthcare applications, there is a plethora of efforts driven by the research and industry communities that aim to overcome the associated challenges and realize the full potential of these technologies toward improving the health, well-being, and independent living of patients and the elderly [34,62,63]. In this context, the following subsections highlight some state of the art research efforts on the topic, presenting three research projects that have recently started or will soon start tackling said issues, each proposing a novel approach and investigating different angles of the IoT/IoMT-enabled smart healthcare landscape. More specifically, the presented projects include SMART BEAR, sustAGE and xVLEPSIS.

2.5.1 SMART BEAR—smart living solution platform for the elderly

The SMART BEAR project aims to provide an intelligent and personalized digital solution for sustaining and extending healthy and independent living by implementing an affordable, accountably secure, and privacy-preserving innovative platform. This system boasts off-the-shelf smart and medical devices to support the healthy and independent living of elderly people with five prevalent health-related conditions: HL, CVDs, CIs, MH issues, and BDs, as well as frailty. This will be achieved through intelligent, evidenced-based interventions on lifestyle, medically significant risk factors, and chronic disease management. These interventions are enabled by the utilization of continuous and objective medical and environment sensing, assistive technologies, and big data analytics.

In more detail, the SMART BEAR platform integrates heterogeneous sensors and assistive devices that collect and analyze data streams from the activities of the elderly with modules that extract the necessary evidence to design personalized interventions to promote healthy and independent living. The platform will also be connected to hospital and other healthcare service systems to obtain data specific to the end users (e.g., medical history) that will need to be considered in making decisions for interventions. SMART BEAR will leverage big data analytics and learning capabilities, allowing for large scale analysis of the abovementioned collected data, to generate the evidence required for making decisions about personalized interventions. Privacy-preserving and secure by design data handling capabilities protect data at rest, in processing, and in transit and will comprehensively cover all the components and connections utilized by the SMART BEAR platform. An overview of the SMART BEAR platform is depicted in Fig. 2.1. To achieve the above, SMART BEAR will build on the platform developed within the H2020 project EVOTION (http://h2020evotion.eu/) to support evidence based public health policies formation and monitoring. The EVOTION platform supports: (1) the continuous collection of medical, physiological, and lifestyle data from heterogeneous resources including hospitals, biosensors, advanced hearing aids, and mobile phones and (2) the analysis

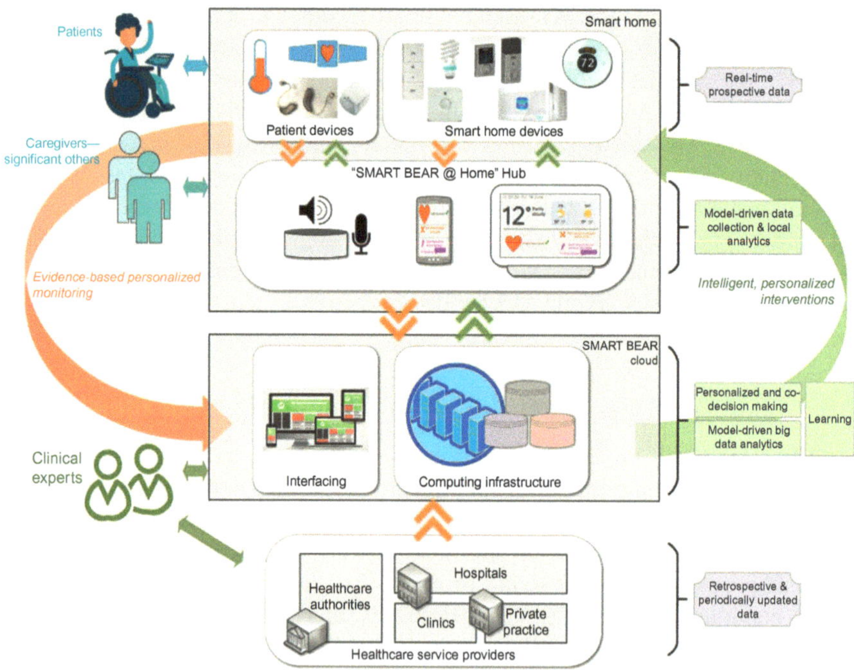

FIGURE 2.1

The SMART BEAR concept.

of these data, driven by high-level big data analytics and decision models to generate evidence useful for making public health policy level interventions [64–66]. The EVOTION platform is currently used in five hospitals in Greece and the United Kingdom, collecting real-time data from more than 1000 hearing aid users.

Key areas of innovation for SMART BEAR will include:

1. integration with IoT enablers and platforms (e.g., FI-WARE, Copernicus, consumer smart ecosystems), in order for SMART BEAR to extend the connectivity of the EVOTION platform to support new medical devices, wearables, smart home/IoT sensors and actuators, and smart environment infrastructures;
2. development of new high-level data analytics and decision models to support the intelligent and personalized interventions required for enhancing the healthy and independent living of the elderly;
3. integration of the EVOTION platform with a continuous security and privacy assurance platform to provide the continuous auditability and transparency needed for ensuring the SMART BEAR platform's trustworthiness by its end users, and
4. testing and validation of the above at a much greater scale, involving 5000 participants across five countries.

In developing the above extensions, special consideration will be given to creating an extensible and sustainable platform, open for wider adoption in the connected health ecosystem.

2.5.1.1 Targeted pilot environments

The SMART BEAR platform will be tested and validated through five large scale pilots, involving 5000 elderly users living at home in Greece, Italy, France, Spain, and Romania. The pilots will enable the evaluation of the platform in the context of healthcare service delivery by private and public providers at the regional, state, and EU levels, and demonstrate its efficacy, extensibility, sustainability, and cost effectiveness for the individual and the healthcare system. SMART BEAR will benefit from this diversity as data coming from all pilots will be collected and evaluated.

More specifically, the Greek pilot will run in two regions with different characteristics in order to evaluate the efficiency of the SMART BEAR solution in different socioeconomic conditions. These will be the Municipality of Palaio Faliro (a metropolitan area with approximately 10,000 people over 65 years of age) and the Region of Peloponnese (a rural area with a significant portion of elderly population). The Italian pilot will cover both rural and urban territories in Lombardy, so as not to restrict the sampling of this pilot to a single geographical area. Two areas are covered by the pilot: the metropolitan area of Milan (8.2 million inhabitants over an area of about 13,000 km^2) and the District of Crema (150,000 inhabitants over an area of about 573 km^2). The two areas are very different because of their extent, environmental conditions, urban services, and population. Concerning the French pilot, two regions are considered as possible and interesting experimentation areas: (1) Île-de-France (the Paris region), the area with the largest number of elderly people (and thus, of dependent elderly people); (2) Nouvelle Aquitaine (particularly the "Creuse" department), the region where the population is the oldest and where many innovative eHealth programs and projects are developed for elderly people, and (3) Bretagne, where the elderly people are the healthiest and which is an innovative and dynamic region in the eHealth field. In the Spanish pilot, the focus will be on the Basque Country, spanning an area of about 7000 km^2 with 2 million inhabitants, and one of the European regions most affected by the aging process. The pilot will cover independent elderly users living at home, seniors living in rural areas, as well as those living in collective structures, such as senior residences. Finally, in the Romanian pilot, participants will come mainly from the capital Bucharest, with a population of about 2 million people, of which 17% are over 65 (359,182). Bucharest is the area with both the largest number of elderly people (three times higher than in any other administrative region of the country) and the largest number of dependent elderly people.

2.5.1.2 The SMART BEAR consortium

SMART BEAR participants collectively constitute a consortium capable of achieving the project objectives, both well-suited and committed to the tasks assigned to them. The SMART BEAR consortium consists of 25 organizations, including four big industry partners in the ICT domain (ATOS Spain S.A. from Spain; Philips Electronics Nederland B.V. from the Netherlands; International Business Machines Corporation from Israel; and Lombardia Informatica from Italy). Moreover, the SMART BEAR consortium includes five partners from the healthcare domain (Comunita' Sociale Cremasca and the Fondazione Centro San Raffaele from Italy; CATEL from France; MUTUALIA from Spain; and Fundatia Ana Aslan International from Romania), as well as two local authorities (Region of Peloponnese and Municipality of Palaio Faliro from Greece). Part of the consortium are also eight large academic/research organizations (CNR ICAR from Italy; Foundation for Research and Technology–Hellas, National Kapodistrian University of Athens, and University of Ioannina from Greece; Università degli Studi di Milano from Italy; Universidad del País Vasco/Euskal Herriko Unibertsitatea from Spain; City, University of London from the United Kingdom, and Institute of Communication & Computer Systems from Greece) as well as six SMEs (Sphynx Technology Solutions AG from Switzerland, StreamVision from France, IT Support Solutions from Romania, Innovatec from Spain, Athens Technology Centre from Greece, and Bird and Bird from the United Kingdom). All these providers bring not only their technological expertise but also their entrepreneurial aspiration regarding their role in creative industries.

2.5.2 sustAGE—smart environments for person-centered sustainable work and well-being

sustAGE[9] is a person-centered smart solution that aims to promote the concept of "sustainable work" for EU industries, thus supporting the well-being, wellness at work, and productivity of aging employees through three main dimensions. The first dimension is directed toward improving occupational safety and health via risk assessment and prevention strategies based on workplace and person-centered health surveillance monitoring. The second dimension aims to promote the well-being of employees via personalized recommendations for physical and MH improvement. The third dimension supports decision-making related to task/job role modifications and aims to optimize overall workforce productivity by assessing the abilities of individual persons (e.g., physical, mental, social) in relation to work demands and risks. The sustAGE solution explores two industry domains with significant challenges and requirements, specifically (1) manufacturing and (2) transportation and logistics.

2.5.2.1 The industry domains

2.5.2.1.1 The case of assembly line workers in the automotive industry

There are hundreds of tasks in the manufacturing assembly process, which differ in terms of posture, workload, and complexity and require both manual labor as well as significant cognitive workload. In the automotive industry, assembly lines can produce two to three different models of a vehicle, each with dozens of possible variations. There is a small tolerance for errors in an often customizable production unit; therefore, workers need to be constantly aware of the specific order and customizations needed to be made. Furthermore, to choose the best match between task and worker in both repetitive short-cycle task operations and complex tasks, worker profiling on an individual and frequent basis is necessary to assess a worker's physical abilities and mental skills. To further take into account age-related changes, it is important to monitor both the environmental conditions and the worker's health state and actions to derive information on the individual's workload that may further impact their physical and mental state. Actions to be monitored are user proximity to critical areas, repetitive movements, bend or twisted postures, or pushing/pulling/lifting an object, along with the temporal aspects of the action (e.g., time to complete the action, pace).

2.5.2.1.2 The case of port workers in the transportation and logistics industry

Port work activities involve loading and unloading procedures and transport and storage of goods (e.g., container movement and roll on/roll off). Pilotage, work-boat and tug operation, ship repairs, vessel traffic management, and similar marine activities are also involved. Dock workers are usually exposed to stressful

[9] http://www.sustage.eu (accessed on June 8, 2019).

and dangerous working conditions. Commonly shift work (morning, afternoon, night) can result in sleep deprivation, misalignment of circadian rhythms, drowsiness, and performance deficits. Noise, vibrations, dust, wind, and tide commonly occur in ports. Workers who perform handwork and require physical strength to carry out activities are prone to musculoskeletal disorders. Beyond the physical extent of port work, the mental demands of attention and concentration at work are important as workers need to be continuously alert. The main case of interest in sustAGE regards the loading/unloading procedures of containers, in which the system monitors the container crane operator and the workers involved in the loading/unloading procedures as well as other moving objects/humans in proximity to the crane during maneuvering. The actions to be detected in this case to support the analysis, profiling, and recommendations of the system are fatigue, as derived by tracked movements along with temporal properties, as well as physiological measurements and proximity of workers to critical areas and moving objects.

2.5.2.2 Internet of Things ecosystem and system functionalities

The developed system functionalities build upon an IoT ecosystem based on off-the-shelf sensors integrated in daily devices and in the work environment, considering both indoor (manufacturing) and outdoor (port) working conditions. The system gathers contextual information from the working environment and from users' physiological signals, tasks, activities, and behavioral patterns, in order to support user profiling and provide personalized recommendations for better managing health, wellness and safety. The sustAGE technology will consider information-rich micromoments[10] (a highly investigated topic of leading technology companies like Google, Microsoft, and Facebook, geared to be the "next big thing" in intelligent system design) to process the short- and long-term aspects of symbiotic interaction, to identify patterns of human behavior, draw correlations between actions, predict what humans do and do not want, improve user's acceptance, and engage users in a successful long-term interaction. Therefore, the notion of time, the consideration of real-world phenomena, and interactions in association with the course of time is very important. Measurements collected from different devices and modules of the system support the definition of key micromoments for future user profile updates, recommendations, and notifications. Different micromoments related to the user daily schedule, work environment, workload, physical/emotional/mental state, and social activities are used (Table 2.2).

Indicative key features of the sustAGE solution are:

- *Monitoring of user actions and behaviors in work environment and personal life*. An IoT ecosystem is exploited, comprised of smart sensors and mobile devices for locating and tracking users in real time and for the fine-grained

[10] https://www.thinkwithgoogle.com/marketing-resources/micro-moments/

Table 2.2 List of indicative micromoments.

Category	Indicative list of micromoments
Work environment	High/low temperature, noise level, pollution, wet/dry weather
Work/task	Work shift, task onset, task completion, task type (repetitive work, bent or twisted body posture), push/pull an object, lifting heavy load, task switch, pace, task break, injury from accident/body part
Daily schedule	Arriving at-leaving from work, lunch/dinner, medication intake, wake-up, go to sleep, meet friends
Physical	Health check, instance of pain, high/low pulse rate, body temperature, walking, resting, steps count during activity, fatigue
Mental	Stress/frustration, depression sign, emotional state changes, state communication/verification by the system
HR	Sick days, tasks increasing/decreasing productivity, employee requests for task changes

detection of user actions and states. By combining information from multiple sources, the system will be able to support user profiling in a privacy-preserving manner and provide context-aware recommendations and analytics.

- *Abstraction and episodic knowledge*. Analyze users' activities and memorize important episodes aiming to keep important information related to past human activities and states. Building on users' micromoments, the system will memorize actions that users need to take and will better predict user reactions by considering their activity in similar past situations.
- *Multiaspect user profiling*. The aggregation of past user-specific knowledge, comprised of user preferences, the results of user performance in work- and training-related activities, and long-term abstractions will allow a more complete physical, mental, and psychosocial user profiling. The collection and analysis of related information will be done transparently, without user intervention.
- *Multilevel personalized recommendations*. Recommendations with respect to three different levels are provided, namely physical, mental, and workforce. Recommendations on the first two will be managed by the individual person measuring the impact on the work ability, health, and well-being, whereas the workforce recommendations will be managed by the management. The system will consider spatiotemporal aspects, taking into account the user's activity, state, time and location, the daily and weekly schedule and will recommend an activity at the right moment.
- *Safe working environment*. Continuously monitor both the environmental conditions in the working area (i.e., manufacturing floor or port dock) and workers' health-related signs in order to detect critical cases and workload issues early and to provide alerts to specific workers, who must take short breaks or switch tasks for the rest of their shift or over longer intervals.

The IoT infrastructure comprises of the following devices/sensors:

- *Environmental sensors* measuring air temperature, humidity, air quality, pressure, dust concentration and noise based on Raspberry Pi/Arduino custom sensors that are open source, low cost, accurate, and durable.
- *Cameras* installed in key working areas. For the manufacturing indoor environment of the assembly line, passive stereo cameras are used to monitor postures and repetitive actions of users, whereas for the port outdoor environment, monocular cameras with varying focal lengths are used to monitor crane operators and workers involved in loading/unloading of containers and people/objects in the vicinity of the crane.
- For localization in indoor environments, *beacons* achieve a precision of up to 10–20 cm within a range of up to 100 m, whereas for outdoor environments the *GNSS receivers* built-in smartphones are used.
- *Wristwatch devices* gather physiological measurements and are able to trigger notifications to users from the system. The selected device should offer software development kits (SDKs) and APIs to facilitate its programmability and access to the data.
- *Smartphone* devices able to support Galileo, offering centimeter accuracy and the ability to communicate with the wristwatch device.

The above set of devices/sensors can collaboratively provide information on different user activities/actions (e.g., walk, bend, stand/sit, push/pull object), state (e.g., fatigue, discomfort), temporal aspects, and specific events in the environment (e.g., user monitoring in specific areas, proximity to hazardous conditions). Moreover, the smartphone is the primary device for communication and multimodal interaction supporting natural language understanding and sentiment analysis. The adopted IoT configuration exhibits the advantages of unobtrusive user-context interaction monitoring in a privacy-preserving way considering that in private life, outside the working environment, only the wristwatch and the mobile device are to be used. The system supports raw data processing near the end-devices to prevent potentially privacy-sensitive information from being sent to the upper layers of the platform in the cloud.

2.5.2.3 The sustAGE consortium

The sustAGE consortium comprises a unique blend of partners from disciplines that span a broad spectrum. The project brings together one of the largest European automotive industries (Centro Ricerche Fiat Scpa, Italy), one of Greece's most important maritime ports (Heraklion Port Authority, Greece), a global leader in ICT products and services (Software AG, Germany), SMEs providing expertise in interactive technologies for e-health (Imaginary Srl., Italy) and distributed systems (AEGIS IT Research UG, Germany), three top European universities in the areas of embedded intelligence for health care and well-being (University of Augsburg, Germany), aging and neurodegenerative diseases (Universidad Nacional de Educación a Distancia, Spain), positioning and sensors (Aristotle University of

Thessaloniki, Greece), two top European research centers in the areas of ergonomics, working environments and human factors (Forschungsgesellschaft für Arbeitsphysiologie und Arbeitsschutz E.V.), and emerging ICT research (Foundation for Research and Technology–Hellas).

2.5.3 xVLEPSIS—an intelligent noninvasive biosignal recording system for infants

Over the last years there is a strong interest in improving patient monitoring in an attempt to facilitate clinicians providing error free decisions while saving time and improving the overall quality of patient care. Such approaches are particularly useful in time critical settings, such as the intensive care unit of the hospital. A stronger effort is required to provide high quality, multimodal, real-time neonate monitoring platforms that can be ubiquitous and unobstructive while at home.

xVLEPSIS[11] is an advanced system for the prediction of potentially hazardous events related to infants. As many pathological situations can occur during an infant's night sleep that can potentially be threatening to health if not detected promptly, there is an imperative for early detection of medical emergencies during infant sleep through an unobtrusive and noninvasive detection system. Invasive devices and sensors could disrupt the infant's sleeping phases that are extremely important for their development and degrade the quality of their rest.

xVLEPSIS uses a scalable system comprising a "smart" bed mattress and a camera positioned to monitor the infant cradle. Without disturbing the infant's sleep, the system can detect possible pathological conditions.

2.5.3.1 Integration of smart biosignal sensors in a detection system for hazardous conditions

The xVLEPSIS system will incorporate diverse user-friendly electronic smart sensors, integrated under a "smart" mattress, in combination with a high-resolution baby monitor. In brief, the following biosignals will be recorded, analyzed, and investigated for their applicability as biomarkers for certain pathologies:

- video recording using a high-resolution camera and audio recording using a high definition microphone;
- ballistocardiogram [67] recording, which records sudden blood ejections into the great vessels with each heartbeat;
- pressure sensors under the bed mat record and plot repetitive body movements during sleep; and
- temperature and humidity detection using suitable sensors under the bed mat.

[11] https://xvlepsis.gr/en (accessed on June 8, 2019).

The development of an intelligent system that will detect potentially hazardous pathological conditions with the use of sophisticated machine learning techniques will lead to:

1. A mobile or smart watch–based notification system, which will alert the parents in the case of emergency.
2. Continuous biosignal recording, throughout the infant's sleep. The recorded data could be sent to the doctor or the hospital, in the case that an abnormality is detected, or they could be evaluated by the doctor during regular infant examination, in the case nothing critical is detected. Therefore, the pediatrician will be able to examine and evaluate all the available medical data and detect any incidents that may have occurred at night without having been perceived by the parents.

Many advantages arise from the development of a low-cost product with all the aforementioned features:

- Continuous recording of high definition video and audio will allow for a more effective monitoring of the infant, whereas the pathological situations detection system will lead to the discovery of incidents that would otherwise remain unnoticed.
- The proposed noninvasive monitoring system will aid the diagnosis and proper treatment of medical disorders that can occur while the parents are not present (e.g., febrile convulsions, epileptic seizures, or apnea).
- Pediatricians always face the challenge of evaluating medical incidents solely based on the information that parents provide, which is not objective and accurate, especially during the first year of infants life. The proposed integrated system offers the medical professionals in charge the opportunity to assess those incidents based on detailed recorded biosignals and, thus, form a better opinion on the diagnosis.
- The use of innovative machine learning algorithms performed on the multimodal medical signals will significantly aid the detection of new quantitative biomarkers of the relevant diseases.
- The medical database that will be implemented will significantly contribute to the research and study of early childhood disorders.

Such a system is expected to effectively notify the parents and enable doctors to identify specific pathologies. The system will continuously and unobtrusively record important biosignals and analyze them using sophisticated machine learning algorithms, suitable for pathology-specific pattern identification and biomarker extraction. The software to be developed will act as a recommendation and alarming system to notify the parents and/or the physician, if needed, by means of a notification center hosted in a smartphone and/or a smartwatch. A dedicated repository will host raw signals for future reference or doctor's referral.

2.6 Conclusion

The role of smart biosensors and IoT is significant in modern medical care. Patients, care-providers and health professionals can strongly benefit from smart applications developed on top of such infrastructures. In this chapter, we presented the challenges and opportunities from the application of smart biosensors in healthcare and described three state of the art solutions that employ smart sensors in this context. The applications demonstrate how smart living solutions can be developed on top of an ecosystem that combines IoT and smart biosensors to record and analyze biosignals in a noninvasive way and allow the early detection and prevention of potentially hazardous pathological conditions. Since there are still many challenges concerning data privacy, data aggregation and integration, and intelligent decision-making to be overcome, this effort has to be intensified. Future efforts should focus on the use of data analysis and data mining techniques as well as the development of machine learning models that can efficiently handle biosignal data streams and effectively decide on the proper actions to take.

Acknowledgments

This work has been partially supported by the European Union's Horizon 2020 Research and Innovation program under Grant Agreements No. 857172 (EU project SMART BEAR) and No. 826506 (EU project sustAGE). This work has also been partially supported by the European Union and Greek national funds through the Operational Program Competitiveness, Entrepreneurship and Innovation, under the call Research–Create–Innovate (project xVLEPSIS, T1EDK-04440).

References

[1] S. Ajami, F. Teimouri, Features and application of wearable biosensors in medical care, J. Res. Med. Sci. 20 (12) (2015) 1208–1215. Available from: https://doi.org/10.4103/1735-1995.172991.

[2] P. Mehrotra, Biosensors and their applications – a review, J. Oral Biol. Craniofacial Res. 6 (2) (2016) 153–159. Available from: https://doi.org/10.1016/j.jobcr.2015.12.002.

[3] WHO. World Report on Ageing and Health, World Health Organization. Available from: <https://apps.who.int/iris/bitstream/handle/10665/186463/9789240694811_eng.pdf?sequence = 1>, 2015 (accessed 08.06.19).

[4] T. Vos, A.A. Abajobir, K.H. Abate, C. Abbafati, K.M. Abbas, F. Abd-Allah, et al., Global, regional, and national incidence, prevalence, and years lived with disability for 328 diseases and injuries for 195 countries, 1990–2016: a systematic analysis for the Global Burden of Disease Study 2016, Lancet 390 (10100) (2017) 1211–1259.

[5] WHO. Cardiovascular Diseases (CVDs) Key Facts, World Health Organization. Available from: <http://www.who.int/en/news-room/fact-sheets/detail/cardiovascular-diseases-(cvds)>, 2017 (accessed 08.06.19).

[6] L. Leto, M. Feola, Cognitive impairment in heart failure patients, J. Geriatr. Cardiol. JGC 11 (4) (2014) 316.

[7] EIPAHA—European Innovation Partnership on Active and Healthy Ageing. A compilation of good practices on prevention and early diagnosis of frailty and functional decline, both physical and cognitive in older people, European Union, Brussels, Belgium. Available from: <https://ec.europa.eu/research/innovation-union/pdf/active-healthy-ageing/gp_a3.pdf>, 2012 (accessed 08.06.19).

[8] K. Murad, D.W. Kitzman, Frailty and multiple comorbidities in the elderly patient with heart failure: implications for management, Heart Fail. Rev. 17 (4–5) (2012) 581–588.

[9] K. Jekel, M. Damian, C. Wattmo, L. Hausner, R. Bullock, P.J. Connelly, et al., Mild cognitive impairment and deficits in instrumental activities of daily living: a systematic review, Alzheimer's Res. Therapy 7 (1) (2015) 17.

[10] J. Chowhan, M. Denton, C. Brookman, S. Davies, F.K. Sayin, I. Zeytinoglu, Work intensification and health outcomes of health sector workers, Personnel Rev. 48 (2) (2019) 342–359.

[11] C. Bhatt, N. Dey, A.S. Ashour (Eds.), Internet of Things and Big Data Technologies for Next Generation Healthcare, Springer, 2017. 978-3-319-49736-5.

[12] M.A. Razzaque, M. Milojevic-Jevric, A. Palade, S. Clarke, Middleware for Internet of Things: a survey, IEEE Int. Things J. 3 (1) (2016) 70–95. Available from: https://doi.org/10.1109/JIOT.2015.2498900.

[13] A.B. Brush, B. Lee, R. Mahajan, S. Agarwal, S. Saroiu, C. Dixon, Home automation in the wild, in: Proceedings of the 2011 Annual Conference on Human Factors in Computing Systems—CHI '11, ACM Press, New York, 2011, p. 2115. ISBN 9781450302289. doi: 10.1145/1978942.1979249.

[14] Y.F. Gomes, D.F.S. Santos, H.O. Almeida, A. Perkusich, Integrating MQTT and ISO/IEEE 11073 for health information sharing in the Internet of Things, in: 2015 IEEE International Conference on Consumer Electronics (ICCE), 2015, pp. 200–201. ISBN 978-1-4799-7543-3. doi: 10.1109/ICCE.2015.7066380.

[15] S.-M. Kim, H.-S. Choi, W.-S. Rhee, IoT home gateway for auto-configuration and management of MQTT devices, in: 2015 IEEE Conference on Wireless Sensors (ICWiSe), 2015, pp. 12–17. ISBN 978-1-4673-9398-0. doi: 10.1109/ICWISE.2015.7380346.

[16] D. Driscoll, A. Mensch, T. Nixon, A. Regnier, Devices profile for web services, version 1.1, 2009. Available from: <http://docs.oasis-open.org/ws-dd/dpws/wsdd-dpws-1.1-spec.pdf> (accessed 08.06.19).

[17] K. Fysarakis, I. Papaefstathiou, C. Manifavas, K. Rantos, O. Sultatos, Policy-based access control for DPWS-enabled ubiquitous devices, in: Proceedings of the 2014 IEEE Emerging Technology and Factory Automation (ETFA), Barcelona, Spain, 2014, pp. 1–8. IEEE. ISBN 978-1-4799-4845-1. doi: 10.1109/ETFA.2014.7005233.

[18] Z. Shelby, K. Hartke, C. Bormann, The constrained application protocol (CoAP). Available from: <https://tools.ietf.org/html/rfc7252>, 2014.

[19] G. Elhayatmy, N. Dey, A.S. Ashour, Internet of Things based wireless body area network in healthcare, Internet of Things and Big Data Analytics Toward Next-Generation Intelligence., Springer, Cham, 2018, pp. 3–20.

[20] A. Jain, B. Vishal, Concoction of ambient intelligence and big data for better patient ministration services, Int. J. Ambient Comput. Intell. 8 (4) (2017) 19–30.

[21] M.A.A. Da Cruz, J.J.P.C. Rodrigues, J. Al-Muhtadi, V.V. Korotaev, V.H.C. de Albuquerque, A reference model for Internet of Things middleware, IEEE Int.

Things J. 5 (2) (2018) 71–883. Available from: https://doi.org/10.1109/JIOT.2018.2796561.

[22] I.C. Jeong, D. Bychkov, P.C. Searson, Wearable devices for precision medicine and health state monitoring, IEEE Trans. Biomed. Eng. 66 (5) (2019) 1242–1258. Available from: https://doi.org/10.1109/TBME.2018.2871638.

[23] D. Dias, J. Paulo Silva Cunha, Wearable health devices-vital sign monitoring, systems and technologies, Sensors 18 (8) (2018) 2414. Available from: https://doi.org/10.3390/s18082414.

[24] Y. Mendelson, D.K. Dao, K.H. Chon, Multi-channel pulse oximetry for wearable physiological monitoring, in: Proceedings of the 2013 IEEE International Conference on Body Sensor Networks (BSN), May 6–9, 2013, MA, 2013, pp. 1–6.

[25] T. Xiao-Fei, Z. Yuan-Ting, C.C.Y. Poon, P. Bonato, Wearable medical systems for p-Health, IEEE Rev. Biomed. Eng. 1 (2008) 62–74. Available from: https://doi.org/10.1109/RBME.2008.2008248.

[26] A. Henriksen, M. Haugen Mikalsen, A.Z. Woldaregay, et al., Using fitness trackers and smartwatches to measure physical activity in research: analysis of consumer wrist-worn wearables, J. Med. Internet. Res. 20 (3) (2018) e110. Available from: https://doi.org/10.2196/jmir.9157.

[27] R. Zangróniz, A. Martínez-Rodrigo, J.M. Pastor, M.T. López, A. Fernández-Caballero, Electrodermal activity sensor for classification of calm/distress condition, Sensors 17 (10) (2017) 2324. Available from: https://doi.org/10.3390/s17102324.

[28] W.Z. Khan, Y. Xiang, M.Y. Aalsalem, Q. Arshad, Mobile phone sensing systems: a survey, IEEE Commun. Surv. Tutor. 15 (1) (2013) 402–427.

[29] Y. Ranjan, M. Kerz, Z. Rashid, S. Böttcher, R.J. Dobson, A.A. Folarin, Radar-base: a novel open source m-health platform, in: Proceedings of the ACM Intel. Joint Conference and 2018 International Symposium on Pervasive and Ubiquitous Computing and Wearable Computers, UbiComp '18, 2018, pp. 223–226.

[30] J. Frankle, M. Carbin, The lottery ticket hypothesis: finding sparse, trainable neural networks. Available from: <https://arxiv.org/abs/1803.03635>, 2018.

[31] A. Kumar, H. Kim, G.P. Hancke, Environmental monitoring systems: a review, IEEE Sensor. J. 13 (4) (2013) 1329–1339. Available from: https://doi.org/10.1109/JSEN.2012.2233469.

[32] G. Mois, S. Folea, T. Sanislav, Analysis of three IoT-based wireless sensors for environmental monitoring, IEEE Trans. Instrum. Meas. 66 (8) (2017) 2056–2064.

[33] E. Kim, S. Helal, D. Cook, Human activity recognition and pattern discovery, IEEE Pervasive Comput. 9 (1) (2010) 48–53.

[34] A.A. Chaaraoui, P. Climent-Pérez, F. Flórez-Revuelta, A review on vision techniques applied to human behaviour analysis for ambient-assisted living, Expert Syst. Appl. 39 (12) (2012) 10873–10888. Available from: https://doi.org/10.1016/j.eswa.2012.03.005.

[35] R. Poppe, A survey on vision-based human action recognition, Image Vision Comput. 28 (6) (2010) 976–990. Available from: https://doi.org/10.1016/j.imavis.2009.11.014.

[36] J. Shotton, T. Sharp, A. Kipman, A. Fitzgibbon, M. Finocchio, A. Blake, et al., Real-time human pose recognition in parts from single depth images, Commun. ACM 56 (1) (2013) 116–124. Available from: https://doi.org/10.1145/2398356.2398381.

[37] M. Sigalas, M. Pateraki, P. Trahanias, Full-body pose tracking – the top view reprojection approach, IEEE Trans. Pattern Anal. Mach. Intell. 38 (8) (2016) 1568–1582.

[38] C. Feichtenhofer, A. Pinz, A. Zisserman, Convolutional two-stream network fusion for video action recognition, in: IEEE Conference on Computer Vision and Pattern Recognition (CVPR), 2016, pp. 1933-1941.
[39] F. Moerchen, Algorithms for time series knowledge mining, Proceedings of the 12th ACM SIGKDD international conference on Knowledge discovery and data mining, ACM, 2006, pp. 668–673.
[40] M. Stacey, C. McGregor, Temporal abstraction in intelligent clinical data analysis: a survey, Artif. Intell. Med. 39 (1) (2007) 1–24.
[41] M. Madkour, D. Benhaddou, C. Tao, Temporal data representation, normalization, extraction, and reasoning: a review from clinical domain, Comput. Meth. Programs Biomed. 128 (2016) 52–68.
[42] W. Raghupathi, V. Raghupathi, Big data analytics in healthcare: promise and potential, Health Inform. Sci. Syst. 2 (2014) 3.
[43] E. Vayena, J. Dzenowagis, J.S. Brownstein, A. Sheikh, Policy implications of big data in the health sector, Bull. World Health Organ. 96 (1) (2018) 66–68.
[44] S. Ilarri, R. Hermoso, R. Trillo-Lado, M.D.C. Rodríguez-Hernández, A review of the role of sensors in mobile context-aware recommendation systems, Int. J. Distrib. Sens. Netw. 11 (11) (2015) 489264.
[45] F. Casino, C. Patsakis, E. Batista, F. Borràs, A. Martínez-Ballesté, Healthy routes in the smart city: a context-aware mobile recommender, IEEE Softw. 34 (6) (2017) 42–47.
[46] Y. Lin, J. Jessurun, B. De Vries, H. Timmermans, Motivate: towards context-aware recommendation mobile system for healthy living, in: 2011 5th IEEE International Conference on Pervasive Computing Technologies for Healthcare (PervasiveHealth) and Workshops, 2011, pp. 250–253.
[47] S. Gupta, S.P. Sood, D.K. Jain, 'Let's exercise': a context aware mobile agent for motivating physical activity, Emerging Research in Computing, Information, Communication and Applications, Springer, New Delhi, 2016, pp. 511–520.
[48] V. Ponce, J.P. Deschamps, L.P. Giroux, F. Salehi, B. Abdulrazak, QueFaire: context-aware in-person social activity recommendation system for active aging, in: International Conference on Smart Homes and Health Telematics, Springer, Cham, 2015, pp. 64–75.
[49] J. Bauer, R.C. Staudemeyer, H.C. Pöhls, A. Fragkiadakis, ECDSA on things: IoT integrity protection in practice, Proceedings of the Information and Communications Security (ICICS 2016), Springer, 2016.
[50] T. Marktscheffel, W. Gottschlich, W. Popp, P. Werli, S.D. Fink, A. Bilzhause, et al., QR code based mutual authentication protocol for Internet of things, in: Proceedings of the IEEE 5th workshop on IoT-SoS: Internet of Things Smart Objects and Services (WOWMOM SOS-IOT 2016), 2016.
[51] R. Hummen, J.H. Ziegeldorf, H. Shafagh, S. Raza, K. Wehrle, Towards viable certificate-based authentication for the Internet of things, in: Proceedings of the 2nd ACM Workshop on Hot Topics on Wireless Network Security and Privacy, ACM, April 2013, pp. 37–42.
[52] M. Bentounsi, S. Benbernou, M.J. Atallah, Privacy-preserving business process outsourcing, in: Proceedings of the IEEE 19th International Conference on Web Services, 2012, pp. 662–663.
[53] C. Gentry, Fully homomorphic encryption using ideal lattices, in: Proceedings of the Forty-First Annual ACM Symposium on Theory of Computing, 2009, pp. 169–178.

[54] M. Li, S. Yu, K. Ren, W. Lou, Y.T. Hou, Toward privacy-assured and searchable cloud data storage services, IEEE Netw. 27((4)) (2013) 56–62.
[55] M. Van Dijk, A. Juels, On the impossibility of cryptography alone for privacy-preserving cloud computing, in: Proceedings of the 5th USENIX Conference on Hot Topics in Security, 2010, pp. 1–8.
[56] B. Parducci, H. Lockhart, E. Rissanen, eXtensible Access Control Markup Language (XACML) Version 3.0, 2013.
[57] A.X. Liu, F. Chen, J. Hwang, T. Xie, Designing fast and scalable XACML policy evaluation engines, IEEE Trans. Comput. 60 (12) (2011) 1802–1817. Available from: https://doi.org/10.1109/TC.2010.274. 2011. ISSN 00189340.
[58] E. Rissanen, "XACML v3.0 Privacy Policy Profile Version 1.0," OASIS. [Online]. Available from: <http://docs.oasis-open.org/xacml/3.0/privacy/v1.0/xacml-3.0-privacy-v1.0.html>, 2015 (accessed 08.06.19).
[59] E. Damiani, G. Gianini, M. Ceci, D. Malerba, Toward IoT-friendly learning models, in: IEEE 38th International Conference on Distributed Computing Systems (ICDCS), 2018, pp. 1284–1289.
[60] Q. Jia, L. Guo, Z. Jin, Y. Fang, Preserving model privacy for machine learning in distributed systems, IEEE Trans. Parallel Distrib. Syst. 29 (8) (2018) 1808–1822. Available from: https://doi.org/10.1109/TPDS.2018.2809624.
[61] M. Yamin, A.A.S. Adnan, Improving privacy and security of user data in location based services, Int. J. Ambient Comput. Intell. (2018) 19–42. 9.1.
[62] M. Haghi, K. Thurow, R. Stoll, Wearable devices in medical internet of things: scientific research and commercially available devices, Healthc. Inform. Res. 23 (1) (2017) 4–15. Available from: https://doi.org/10.4258/hir.2017.23.1.4.
[63] N. Zhu, et al., Bridging e-Health and the Internet of Things: The SPHERE project, IEEE Intell. Syst. 30 (4) (2015) 39–46. Available from: https://doi.org/10.1109/MIS.2015.57.
[64] M. Anisetti, C. Ardagna, V. Bellandi, M. Cremonini, F. Frati, E. Damiani, Privacy-aware big data analytics as a service for public health policies in smart cities, Sustain. Cities Soc. 39 (2018) 68–77.
[65] B. Ye, I. Basdekis, M. Smyrlis, G. Spanoudakis, K. Koloutsou, A big data repository and architecture for managing hearing loss related data, in: 2018 IEEE EMBS International Conference on Biomedical & Health Informatics (BHI), IEEE, March 2018, pp. 174–177.
[66] M. Prasinos, G. Spanoudakis, D. Koutsouris, Towards a model-driven platform for evidence based public health policy making, in: Proceedings of the 29th International Conference on Software Engineering and Knowledge Engineering, 2017.
[67] L. Giovangrandi, O.T. Inan, R.M. Wiard, M. Etemadi, G.T.A. Kovacs, Ballistocardiography – a method worth revisiting, Proc. IEEE Eng. Med. Biol. Soc. (2011) 4279–4282.

Further reading

E.A. Carroll, M. Czerwinski, A. Roseway, A. Kapoor, P. Johns, K. Rowan, et al., Food and mood: just-in-time support for emotional eating, in: 2013 IEEE Humaine Association Conference on Affective Computing and Intelligent Interaction, 2013, pp. 252–257.
ENISA, Security and Resilience in eHealth Infrastructures and Services, 2015.
S. Tamane, K.S. Vijender, D. Nilanjan (Eds.), Privacy and Security Policies in Big Data, IGI Global, 2017.

CHAPTER

3

Wearable electroencephalography technologies for brain–computer interfacing

S.R. Liyanage[1] and Chintan Bhatt[2]

[1]*University of Kelaniya, Kelaniya, Sri Lanka*

[2]*Charotar University of Science and Technology, Anand, India*

Chapter Outline

3.1 Introduction

The recent development of brain–computer interfaces (BCIs), which are devices that use brain signals as a nonmuscular communication channel [1] has provided an important element for the creation of systems capable of brain-to-brain (B2B) communication and precise brain stimulation. These advances have enabled production of noninvasive computer–brain interfaces (CBI). BCI and CBI technologies can be combined to realize the vision of hyperinteraction, noninvasive, computer-mediated B2B communication between subjects.

In B2B systems, the conscious transmission of information between human brains can be achieved through the intact scalp and without intervention of motor or peripheral sensory systems [2]. Pseudorandom binary streams encoding words have been transmitted between the minds of an emitter subject and a receiver subject separated by great distances. Grau et al. [2] have established that Internet-mediated B2B communication is possible via combination of a BCI based on

Wearable and Implantable Medical Devices. DOI: https://doi.org/10.1016/B978-0-12-815369-7.00003-3

voluntary motor imagery-controlled electroencephalography (EEG) with a CBI that induces the conscious perception of phosphenes (light flashes) through neuro-navigated, robotic, transcranial magnetic stimulation. Special care must be taken to block sensory (tactile, visual, or auditory) cues. This early evidence provides critical proof of principle for the development of conscious B2B communication technologies. More fully developed, related implementations will open new research venues in cognitive, social, and clinical neuroscience as well as the scientific study of consciousness. We envision that hyperinteraction technologies will have a profound impact on the social structure of our civilization and their development raises important ethical issues.

In this article, the current state of EEG-based BCI systems used for communication is reviewed with a critical analysis of the future possibility of utilizing wearable technologies to facilitate the realization of hyperinteraction.

3.2 Current state of brain–computer interface-based communicators

The potential of utilizing EEG signals for brain–computer interfacing was first demonstrated by Grey Walter in 1964 [3]. The term *brain–computer interface* was apparently first used by Jacques Vidal in 1973 [4]. Modern neuroscience is founded on the sensorimotor hypothesis of brain function [5]. The sensorimotor hypothesis states that the main function of the central nervous system (CNS) is to respond to external and internal events with appropriate outputs. BCIs are defined as systems that bypass the natural neural pathways for these outputs and translate brain signals into new kinds of outputs.

It is estimated that one out of 3000 individuals in the World suffer from some type of stroke syndrome, motor neuron disease, or muscular dystrophy. These patients can gradually reach locked-in state where normal communication through the CNS is lost. It has been reported that there are approximately 25,000 locked-in patients in the United States alone [6]. Locked-in patients are conscious but cannot control any muscle (except for the eyes) or communicate verbally with others. BCI-based communication is the most effective way to facilitate communication with locked-in patients [7]. BCI aims to build a direct channel between the human brain and the computer. It provides an alternative communication and control channel to convey messages and commands from the brain to the external world.

Physiotherapy, occupational therapy, speech therapy, and language therapy are usually administered for rehabilitation of locked-in stroke patients. Modern rehabilitation techniques attempt to resuscitate injured brain areas and to induce non-injured brain areas to take over the functions previously executed by the injured areas. This is achieved by identifying the level of activity-induced plasticity of the brain [8,9]. BCIs can be employed to induce the return of brain activity needed to enable voluntary control of the limbs and/or strengthen existing functional neural pathways by pairing movement intentions with actual movements. Factors such as the amount of time that can be spent in therapy clinics and the number of visits that are covered by insurance impact the effectiveness of rehabilitation therapies [10]. Therefore, BCIs suitable for home use without the need for a therapist or technical assistance may enable far more effective and faster rehabilitation. Mane et al. [11] have shown the effectiveness of EEG-based BCI in rehabilitating stroke patients undergoing different types of motor rehabilitation interventions.

There are two primary types of BCIs, invasive and noninvasive. Invasive BCIs provide precise control but require electrodes to be implanted on the brain surface to capture neuron signals. Noninvasive BCIs use electrodes placed on the scalp and have wider applications due to their convenience and safety. Currently, EEG is the most popular brain signal for noninvasive BCIs, because it is the least expensive compared to other methods for brain signal acquisition such as, functional magnetic resonance imaging (fMRI) and near-infrared spectroscopy (NIRS). A general block diagram of an EEG-based BCI is depicted in Fig. 3.1, and a real application is shown in Fig. 3.2. Electrodes are placed on the surface of the patient's scalp and acquire EEG signals from brain. After signal amplification and digitization, EEG signals are sent to a computer for feature extraction and pattern recognition using digital signal processing software. When a certain pattern is recognized, its corresponding command will be executed and the results displayed on a computer monitor. The user can then make further reactions based on the results.

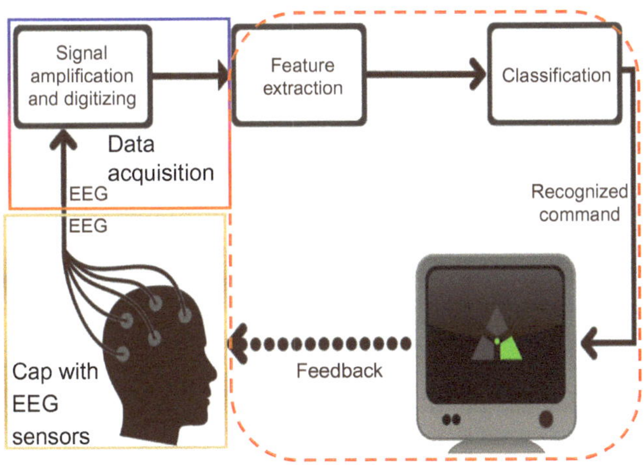

FIGURE 3.1

General block diagram of an electroencephalography (EEG)-based brain–computer interface (BCI).

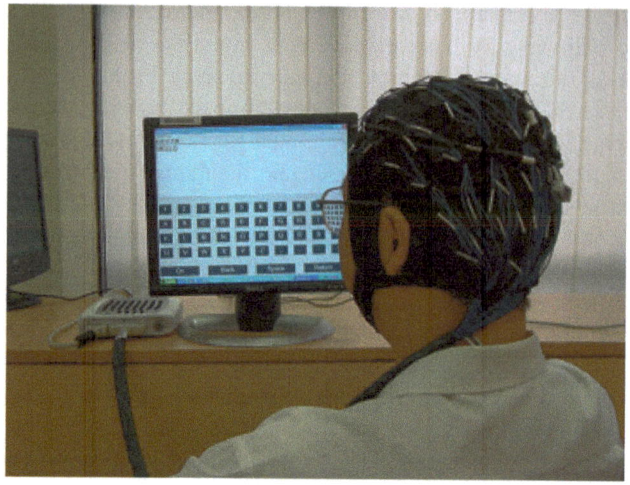

FIGURE 3.2

A real application.

The human brain is one of the most complex systems known. Accordingly, EEG signals are extremely complex and dynamic, vary across individuals, and many are not well understood. EEG signals also tend to be rather weak and subject to severe noise corruptions. All of these factors pose considerable challenges for BCI technology. Although BCI has existed for over 40 years, significant advancements have been made only in the past 10 years. Researchers discovered an electrical signal that can be recorded from the scalp in response to a certain event (stimulus). The most useful feature of this event-related signal is a positive deflection in voltage, known as the P300 wave, that occurs roughly 300 msafter the stimulus. The presence, magnitude, and timing of the P300 wave can be used in decision-making processes. Research has reported that P300-based spelling systems are able to achieve accuracy of over 90% [12].

The P300 evoked potential is a positive deflection of the EEG that occurs about 300 ms after a specific stimulus that satisfies the oddball paradigm. It was first applied by Sutton et al. [13]. The P300 is usually largest over the central parietal scalp and attenuates gradually as distance from this area increases. P300 was first used as the basis for a BCI by Farwell and Donchin [7]. Current P300-based BCIs allow users to select items displayed on a computer screen. P300-based BCIs have become popular because they can be easily parameterized for a new user, require almost no user training, and are usable by 90% of the population. P300-based BCIs are the most user-friendly BCI-based assistive technology for unsupervised usage by patients with disabilities.

EEG is the primary noninvasive BCI method. Magnetoencephalography (MEG) may also be used. Both EEG and MEG have comparable spatial resolution. The main difference between these methods is their sensitivity to different brain sources. EEG emphasizes synchronized dipole sources in superficial gyral surfaces and larger dipole layers that generate larger scalp potentials. In contrast, MEG emphasizes sources in the sulcal walls, which are tangential to the scalp and thus also tangential to the coils. EEG recording requires careful attention to electrode application, number, and location, as well as to sampling rate, reference choice, and contamination by nonbrain artifacts. Mild skin abrasion and conductive gel are generally needed to obtain adequately low electrode impedance. Dry electrode technology is under development. To avoid temporal aliasing of the EEG, the sampling rate should exceed the Engineer's Nyquist Criterion, that is, it should be at least 2.5 times the highest frequency to which the recording amplifier is sensitive. Similarly, electrode density on the scalp should be high enough to prevent spatial aliasing of the relevant EEG features. The recognition of nonbrain artifacts such as electromyography (EMG) requires adequately comprehensive spatial and frequency analyses. Reference selection is critical. An ideal, completely inactive reference site is not obtainable. Nevertheless, with many electrodes (128 or more) distributed widely over the scalp and use of the average reference electrode method (also called the common-average reference), it is possible to closely approximate reference-independent EEG recording. Although this approach may be ideal for research or clinical purposes, it is impractical for BCIs intended for long-term use in daily life. In such situations a very small number of electrodes (e.g., five), portability, ease of use, and capacity for rapid online processing are highly desirable or even essential. Thus, because the EEG recording in practical BCI systems is necessarily bipolar, it is probably best to incorporate a variety of promising bipolar strategies into BCI research. In this regard, high-resolution EEG methods such as the surface Laplacian technique have potential advantages, especially when a specific cortical area (e.g., motor cortex) produces the signals used by the BCI (e.g., McFarland et al., 2008).

Recent advances in signal acquisition technologies, and advances in signal processing algorithms with the power of machine learning has made BCI a popular topic in research. Currently, most BCI devices are operated in laboratory environments and a few real-world applications, such as gaming, where accuracy and robustness are not critical. For example, in the game "Mindball" [14] developed and commercialized by Interactive Productline, players compete to control the movements of a ball across a table by becoming more relaxed and focused. Similar gaming applications have been developed by Emotiv Systems [15] and NeuroSky [16] in 2009. In contrast, the project proposed here is a medical application of BCI that will provide a communication product for locked-in patients.

EEG signals collected from the scalp can be recorded using two different modes: unipolar and bipolar. In the unipolar mode, channels are formed by electrode-reference pairs where voltage deviations between all electrodes and a reference electrode are recorded. In the bipolar mode, voltage deviations between two specific electrodes are recorded, with each pair forming a channel. The International 10–20 system recommended by the International Federation of

Societies for Electroencephalography and Clinical Neurophysiology (IFSECN) specifies how the electrodes should be placed on the scalp [17].

As the EEG signals collected from different subjects vary, the most suitable channels will have to be calculated individually for best performance. However, with modern headsets that have 64 or more channels, it is important to apply a suitable channel selection algorithm to minimize the computational burden. Signal processing tools such as time domain analysis, power spectral estimation, and wavelet transform have been used for feature extraction and channel selection. Different evaluation approaches such as filtering, wrapping, embedding, and human-based techniques have been widely used for evaluation of the selected subset of channels [18].

The world's first P300-based spelling system, IntendiX, came to the market in 2010 [19]. It works with eight dry active EEG sensors and uses the P300 principle to type on a keyboard-like matrix. A spelling rate of one or two commands per minute can be achieved by the majority of healthy users in their first trial.

Typing rates for current EEG BCI-based communication systems are approximately five characters per minute or one 5-letter word per minute (wpm). People without disabilities typically speak at a rate of 150–250 wpm [20]. BCI can be more difficult to learn than movement-based assistive communication methods, as one must learn not only a new computer interface but also how to control brain activity. In some studies, subjects with disabilities have been found to achieve lower levels of accuracy with BCI than healthy subjects [21,22].

Most BCI systems use bulky, wired laboratory-oriented amplifiers to measure brain activity under controlled laboratory conditions. Such bulky sensing equipment is tedious for users and impedes their ability to perform routine tasks in day-to-day environments. Furthermore, owing to large data volumes, signal processing in BCI systems is often performed offline using high-end personal computers, hindering the applications of BCI in real-world environments. To be practical for routine use by unconstrained, freely moving users, BCI systems must be noninvasive, nonintrusive, lightweight, and capable of online signal processing. This work reviews recent online BCI systems, focusing especially on wearable, wireless, and real-time systems.

The information transfer rate (ITR) of current BCI systems remains a major hurdle for successful integration of BCIs into society [23]. Slow ITR is only one of multiple grand challenges identified for BCI [24]. Other requirements to make BCI effective include advances in data acquisition and interpretation.

Currently, diverse methods are used to monitor brain activity at different levels. These range from highly invasive methods requiring surgery to noninvasive methods that can be worn on the head without any preparation. The signals collected provide only partial information about the subject's intentions. The changing nature of brain activities further complicates monitoring on top of technical problems such as unstable recordings. Therefore, BCI systems must be made adaptive and capable of learning the relevant dynamics of neural processing.

Future BCIs for direct brain communication should rely on strategies requiring minimal or no attention/cognitive effort and should use primarily implicit learning in order to facilitate effective interfacing. Telemetry techniques where power is supplied wirelessly through the use of inductively coupled coils have been attempted to overcome the limitations of current EEG headsets [25].

Application of nanotechnology has the potential to enable development of nanodetectors that can be implanted inertly into the brain to measure neuronal activity. Other technological challenges to using noninvasive techniques involve increasing spatial and temporal resolution and miniaturization of the devices [24].

3.3 Current developments in sensor technology

There are many wearable sensors such as Neurosky, ABM-B-Alert, Quasar, Neuroelectric's Enobio, SmartCap Nia2, Melon, Zeo, and Emotiv Epoc neuroheadset as described in [26]. In light of interest in wireless body area networks for health monitoring, sensors in implanted devices have been integrated with external devices. According to Lin et al. [27], wearable EEG systems capable of identifying drowsiness and arousing the user by giving warning tones have been developed recently.

Sensors fabricated with flexible materials have been attached to persons along with the embedded system needed to monitor a parameter and transfer significant data to the monitoring unit for further analysis. The use of wearable sensors has played an important role in monitoring physiological parameters of individuals in order to minimize the effects of aberrant processes occurring in the body.

Wearable flexible sensors (WFS) have received intense attention in recent years in the attempt to develop sensors with better performance in terms of sensitivity and sustainability. Further research is needed to improve fabrication and implementation of WFS for ubiquitous monitoring. Handling and storage of the massive amount of data generated by the sensors in a wearable system is a critical challenge. Selecting significant information from the massive database for future analysis is also a challenge that needs to be addressed with big data analytic technologies. Security is another area that needs improvement due to the enormous amount of sensitive data generated by WFS systems and the associated need to mitigate mishandling and misuse of collected data.

Time-varying traffic is an additional issue arising during data transmission from different sensor nodes in a real-time topological system. This traffic causes a delay in data reception in the monitoring unit, thus decreasing the efficiency of the system. Some significant data may also be lost due to the high traffic generated by Active Assisted Living applications. Data transmission for a central coordinator system in wireless sensor networks (WSNs) should be handled properly to minimize traffic and loss of data. The connectivity and interoperability of the embedded system should be significant to reduce power and data loss. From a patient's point of view, the user should not experience discomfort due to wearing the WFS. There should not be any breach of privacy for the patient regarding the collected data. The embedded system attached to the sensor should tightly attach to the body or clothes worn by the person to avoid alteration of the data due to movement and the surrounding environment. There is also a risk of thermal effects on the attached sensors caused by the tissues of the patient. Multiple factors contribute to these potential thermal effects [28]. The number of sensors used in the embedded system should be kept as minimal as possible. The location of the sensor is also important. A sensor positioned in the arm will have more thermal effects than a sensor in the chest. The operating frequency of the sensor and network protocol should be as low as possible. Power consumption by WFS is another significant issue that needs to be addressed. Sensors like SHIMMER or Telos with low power consumption should be considered for monitoring purposes to reduce the overall power consumed by the WFS. Continuous supply of power to the system is another challenge that needs to be addressed for future systems. The system should be designed for on-node processing and reduce the effects of motion artifacts and distributed interference.

Printed electronics [29] is another sector that can be utilized for developing future wearable flexible devices [30]. It has always been a challenge to manufacture compatible printed devices with high throughput. Because reduction of the production cost of the sensor is a critical goal, the use of abundant cheap materials to develop intelligent, smart sensors by simple printing processes is intriguing [31]. Some other factors to be considered while developing printed electronics include scalability, environmental friendliness, and mechanical enhancement of devices. The mass production of low-cost materials like plastics and organic substrates would also lead to a wider range of applications [32]. For example, the concept of quantum dots, where semiconducting nanocrystals are tuned for the emission of light based on their resonating wavelength, has been exploited to develop 3D printed light-emitting diodes [33]. The use of printed electronics as wearable sensors has been conceptualized for some time [34]. Because of their high malleability, these printed devices can be easily attached to skin or textiles for monitoring purposes.

The growth in microelectromechanical systems (MEMS) along with nanoelectromechanical systems (NEMS) technology is expected to reduce the cost of fabrication of flexible sensing systems, leading to a wider range of applications in the near future. Utilization of existing manufacturing techniques along with upcoming techniques will assist in developing new sensing systems that will support a better quality of life in the near future.

Advances in EEG sensors have led to small, noninvasive, calibration-less BCI system hardware with advanced machine learning capabilities to overcome problems such as calibration between sessions and subjects. Adaptive signal processing can be used to preprocess brain signals. Asynchronous BCI switching techniques can be applied to switch between the brain–brain interface and BCI system modes as required. Applications in communication, assistive surveillance, and monitoring increase the capacity of the brain being utilized and promote well-being. Wireless wearable BCI systems will significantly reduce setup time and the hassle with wires encountered with wired BCI systems. A wireless BCI system will also provide users with more freedom of movement to complete daily activities outside of a laboratory environment. The user interface of a BCI system can be integrated into daily wearable accessories such as a hat, headband, or pair of spectacles to increase portability [27].

3.4 Current developments in wearable and wireless brain–computer interface

The wearable and wireless BCI system developed and published by Lin et al. [35] featured dry, MEMS EEG sensors, low power signal acquisition, amplification, and digitization, wireless telemetry, and online signal processing.

EEG signals obtained using dry EEG sensors that do not require skin preparation and gel application to ensure good electrical conductivity between the sensor and skin can be used. The acquired signals are first amplified by the signal acquisition and amplification unit. The wireless data transmission unit is comprised of an A/D converter, a complex programmable logic device, and a wireless module. The wireless module could be either an RF module (RF3100/3105) or a Bluetooth module depending on transmission range requirements. The custom-designed DSP module comprising a Bluetooth receiver, data processing, and data display circuits serve as a development platform for general purpose EEG signal processing. The BCI system serves as a platform that can be programmed and extended to many other applications. Persistent BCIs include text spellers with steady state, visually evoked potentials and event-related potentials with viable operation over periods of weeks, demonstrating important advances over alternative BCI technologies. Recent advances in electrodes for noninvasive recording of EEG expand opportunities for collecting such data for BCIs. Existing technologies cannot be used effectively in continuous, uninterrupted modes for more than a few days due to irritation and irreversible degradation in the electrical and mechanical properties of the skin interface. Here we introduce a soft, foldable collection of electrodes in open, fractal mesh geometries that can mount directly on the complex surface topology of the auricle and the mastoid to provide high fidelity and long-term capture of EEGs in ways that avoid any significant thermal, electrical, or mechanical loading of the skin. Experimental and computational studies have established the fundamental aspects of the bending and stretching mechanics that enable this type of intimate integration on the highly irregular and textured surfaces of the auricle [36].

A tongue drive system (TDS) has been proposed by Huo et al. [37], which is an unobtrusive, minimally invasive, wearable, and wireless tongue–computer interface. The TDS can infer its users' intentions, represented in their volitional tongue movements, by detecting the position of a small permanent magnetic tracer attached to the users' tongues. Any specific tongue movements can be translated into user-defined commands and used to access and control various devices in the users' environment. The latest external TDS (eTDS) prototype is built on a wireless headphone and interfaced to a laptop PC and a powered wheelchair. Using customized sensor signal processing algorithms and a graphical user interface, the eTDS performance was evaluated by 13 naive subjects with high-level spinal cord injuries (C2–C5) at the Shepherd Center in Atlanta, GA. Results of the human trial has achieved an average ITR of 95 bits/min with 82% accuracy. This ITR is about two times higher than the EEG-based BCIs that have been tested on human subjects. It was also demonstrated that the subjects had immediate and full control over powered wheelchairs to the extent that they were able to perform complex wheelchair navigation tasks, such as driving through an obstacle course. BCIs can provide nonmuscular communication and control for people with severe motor disabilities. Current BCIs use a variety of invasive and noninvasive methods to record brain signals and a variety of signal processing methods for data analysis. Whatever the recording and processing methods used, BCI performance (e.g., the ability of a BCI to control movement of a computer cursor) is highly variable and, by the standards applied to neuromuscular control, could be described as ataxic. In an effort to understand this imperfection, this paper discusses the relevance of two principles that underlie the brain's normal motor outputs. The first principle is that motor outputs are normally produced by the combined activity of many CNS areas, from the cortex to the spinal cord. Together, these areas produce appropriate control of the spinal motor neurons that activate muscles. The second principle is that the acquisition and lifelong preservation of motor skills depends on continual adaptive plasticity throughout the CNS. This plasticity optimizes the control of spinal motor neurons. In light of these two principles, a BCI may be viewed as a system that converts the outcome of CNS activity from control of spinal motor neurons to control of the cortical (or other) area whose signals are used by the BCI to determine the user's intent. In essence, a BCI attempts to assign to cortical neurons the role normally performed by spinal motor neurons. Thus, a BCI requires that the many CNS areas involved in producing normal motor actions change their roles so as to optimize the control of cortical neurons rather than spinal motor neurons. The disconcerting variability of BCI performance may stem in large part from the challenge presented by the need for this unnatural adaptation. This difficulty might be reduced, and BCI development might thereby benefit, by adopting a "goal-selection" rather than a "process-control" strategy. In "process control," a BCI manages all the intricate high-speed interactions involved in movement. In "goal selection," by contrast, the BCI simply communicates the user's goal to software that handles the high-speed interactions needed to achieve the goal. Not only is "goal selection" less demanding, but in delegating lower level aspects of motor control to another structure (rather than requiring that the cortex do everything), it also more closely resembles the distributed operation characteristic of normal motor control.

Fabric-based electrodes may offer an alternative that is wearable in normal settings, enabling long-term use. We describe preliminary tests to measure and characterize EEG signals using novel fabric-based electrodes produced via seamless knitting and integrated into a comfortable headband. EEG signals were measured from five locations on the forehead, corresponding to locations in the international 10–20 system. Preliminary offline results demonstrated decreased alpha band power (7.5–12 Hz) while performing mental math compared to relaxation (both with eyes closed) in two participants. In further tests, we will investigate the ability to distinguish between mental tasks in real time. A fabric-based EEG headband could significantly impact the lives of individuals with communication disabilities by allowing an unobtrusive and comfortable means of communication through automated sensing of brain activity.

A BCI provides a direct communication channel to the brain without the need for muscles and nerves. With the emergence of wearable and wireless BCIs, these systems have evolved to become part of wireless body area networks, offering people-centric applications such as cognitive workload assessment and detection of selective attention. Currently, wireless body area networks are mostly integrated by low-cost devices that, due to their limited hardware resources, cannot generate secure random numbers for encryption. This is a critical issue in the context of new Internet of Things device communication and its security. Such devices require secure communication, mostly by means of the automatic renewal of the cryptographic keys. In the domain of the people-centric Internet of Things, we propose to use wireless BCIs as a secure source of entropy, based on neuroactivity, capable of generating secure keys that outperforms other generation methods. In our approach current wireless BCI technology is an attractive option for offering novel services emerging from novel necessities in the context of the people-centric Internet of Things. Our proposal is an implementation of the human-in-the-loop paradigm, in which devices and humans indistinctly request and offer services to each other for mutual benefit.

Several alternative wet electrode designs have appeared in recent years. The Emotiv Epoc system uses moistened felt pads and a semirigid support that enables faster electrode placement but is less accurate than conventional placement methods and largely restricts placements to sites on the scalp perimeter, which are more susceptible to EMG contamination [38]. The g.SAHARA dry electrode (g.tec) consists of a set of eight gold-plated pins; these electrodes are mounted in a conventional cap that does not limit electrode locations and is reported to provide P300-based BCI results similar to those provide by wet electrodes [39]. Both the Epoc and g.SAHARA electrodes rely on low impedance resistive contact with the scalp. In contrast, the dry electrode developed by QUASAR and Wearable Sensing uses a hybrid combination of high impedance resistive and capacitive contact to the scalp. BCI methods that use epidural or subdural electrodes [40] or intracortical microelectrodes offer more secure placement and better spatial resolution than EEG. On the other hand, it is not yet clear to what extent and for what purposes their invasive nature and increased cost is justified, as noninvasive methods can produce comparable target acquisition times. In addition, these invasive recording methods, particularly those using intracortical electrodes, have not yet demonstrated reliable long-term recording stability [41]. Future advances in materials and techniques may achieve long-term stability and performance sufficient to justify implanted BCI systems [42].

A sheet-type wireless (EEG) sensor system using flexible and stretchable electronics has been successfully developed on flexible substrates. The EEG sensor system can be attached to the forehead using biocompatible gel with stretchable electronics [43]. Mikkelsen et al. [44] have proposed the keyhole hypothesis where measurements from low dimensional EEG, such as ear-EEG, reflect a broadly distributed set of neural processes. Experimental investigation-based effects of stimuli, such as alpha-attenuation, auditory onset, and mismatch-negativity responses has been conducted for 13 hours. The investigators claim that information acquired at scalp electrodes and through the ear-EEG "keyhole" are consistent. They also claim that large portions of the scalp EEG can be predicted from ear-EEG, with the highest predictability achieved in the temporal regions and when using ear-EEG electrodes with a common reference electrode [44]. With this technique simple single or dual electrodes on the two earlobes might be enough to communicate via BCI with the proper signal processing techniques. Lin et al. [45] proposes a wireless, noninvasive and multifunctional assistive system that integrates steady-state visually evoked potential (SSVEP)-based BCI and a robotic arm to assist patients in feeding themselves. Patients are able to control the robotic arm via the BCI to serve themselves food. Three other functions: video entertainment, video calling, and active interaction are also integrated. This is achieved by designing a functional menu and integrating multiple subsystems. A refined decision-making mechanism is incorporated to ensure the accuracy and applicability of the system. Fifteen participants were recruited to validate the usability and performance of the system. The averaged accuracy and ITR achieved is 90.91% and 24.94 bit/min, respectively. The feedback from the participants demonstrates that this assistive system is able to significantly improve the quality of daily life. Wearable EEG systems that provide complementary recordings have some limitations, such as discomfort, inconvenience of retrieving recordings, inability to provide timely, accurate classification, and early warning information. A recent development using wireless wearable EEG by Zhang et al. [46,47] focused on the following three points: (1) the monitoring device integrated with EEG acquisition sensors, signal processing chip, and Bluetooth, attached to a sport hat or elastic headband; (2) the mobile terminal with dedicated application, which is not only for continuous recording and displaying EEG signals but also for early warning in real time; and (3) the encephalopathy's classification algorithm based on intelligent Support Vector Machine, which is used in a new application of wearable EEG for encephalopathy daily monitoring. Very impressive results have been reported by this system that can be used for effective daily monitoring of EEG and patients' health self-management. Chiu et al. [48] have developed a noninvasive, wireless, SSVEP-based BCI eating assistive system that enables users with physical disabilities to have meals independently by integrating an SSVEP-based BCI system and a robotic eating assistive system. This system has been tested on an amyotrophic lateral sclerosis patients and 15 healthy subjects. An average accuracy of 91.35% was reported by the system.

Liquid metal technologies have made breakthroughs in wearable technologies allowing for the implementation of liquid metal-enabled wearable monitoring and therapy devices [49]. These wearables have advantages such as miniaturization of electronic circuits, data analysis techniques, telecommunication strategies, and physical therapy, etc.

Owing to their portability and continuous action, wearable devices are regarded as promising tools for management and rehabilitation of patients with chronic diseases [50]. Due to the recent developments in wearable electronics and the internet of Things, complex physiological signals can be registered including cerebral (EEG, fMRI, etc.) and peripheral heart rate, biological activity, temperature, etc. [51,52]. These signals can be recorded and processed during various multimodal human–machine interactions. Quantitative characterization and interpretation of the mentioned physiological signals is a nontrivial task that attracts attention of experts from various scientific fields including medicine, biology, chemistry, electrical engineering, computer science, etc. [53,54].

Sensors and miniaturized devices enhance e-healthcare facilities in many ways. Healthcare units use them either inside or outside of the human body. In most e-healthcare applications, smart sensors that absorb (receive), store, and transmit data can be utilized in many different cases, such as monitoring patients' continuous fitness. Thayananthan and Basuhail studied the integration of wearable sensors for monitoring patients [55]. Miniaturized medical devices based on nanotechnology and deoxyribonucleic acid computers provide implantable options. In e-healthcare applications, miniaturization is an innovative technology because it allows healthcare units to integrate wearable sensors. Further, medical devices with smaller, superior features and capabilities inspire the next generation of the e-healthcare systems. According to [56], applications of a system depend on an efficient system architecture that includes the latest key technologies, such as integration with the Internet of Things, large health systems, and human-cloud integration [57,58]. Design and performance analysis of wireless body sensor networks allow patients to improve the healthcare problems at different levels through Android [59].

Novel EEG acquisition hardware that are able to record EEG signals without affecting the user's performance of other tasks are available today [60]. Wearable EEG systems can be found in a range of shapes as well. Some are incorporated into headbands [60–62], while others are inside headphones [63,64], caps [65,66], helmets [67], or headsets [68,69]. The usability of some of the available systems has been reviewed in [70].

Whitchurch et al. [71] first tried wireless BCI with a Bluetooth based system. Obeid et al. [72] were able to successfully implement a 16-channel BCI system through a WLAN with the IEEE802.11b transmission protocol. Matthews [67]

Table 3.1 Comparison of wireless BCI systems.

System	Signal	Channels	Transmission	Gain	Sampling rate	Filter	Signal processor
[71]	EEG	6	Bluetooth	10,000	–	60 Hz low pass	Notebook computer
[72]	Single Unit	16	WLAN IEEE802.11b	500/ 1000	244	211 Hz high pass	66 MHz AMD
[67]	EEG	Multiple channel	–	–	–	–	Computer/ PDA
[73]	EEG	Multiple channel	Bluetooth	–	500	–	Computer
[74]	EEG + fNIRS	24 for fNIRS 37 for EEG	Multimodal	–	–	–	Computer

and Cincotti et al. [73] also attempted wireless BCI with multiple channels. Hybrid BCI where EEG is complemented with another signal modality such as NIRS, EMG, etc., has become popular due to increased accuracy. In this context, combined, synchronous measurements of functional near-infrared spectroscopy (fNIRS) and EEG has been successfully attempted in [74]. Table 3.1 summarizes some of the EEG acquisition devices that are capable of wirelessly collecting EEG signals and transmission of data.

In order to make wearable BCI systems a reality, further research is needed in both hardware and software. Advances in EEG hardware are quite promising. Dry-contact and noncontact-type EEG sensors have been developed to improve EEG measurements and to overcome problems related to wet electrodes such as time-consuming skin preparation and the need to apply gel or saline liquid regularly [61,62]. Fabric-based sensors were proposed for measuring biopotential signals [75–78]. Foam-based sensors have also demonstrated reduced motion artifacts when obtaining measurements from a user using conventional wet electrodes while moving [79]. Novel semidry electrodes embedded in a headset for EEG recording, aiming to overcome the hurdles of current wet and dry electrodes, have been investigated in [80]. Semidry electrodes have been found to be able to release a small amount of saline solution with assistance of capillary force through porous ceramic pillars, eliminating the need for skin preparation and gel application. The semidry electrodes have demonstrated many advantages including quick setup, self-application, up to 8-hour usage window and cleanliness for users [80].

Fabric-based sensors are relatively comfortable and noninvasive. However, fabric-based sensors for acquiring EEG signals at sites covered with hair require further improvement because the contact area of the skin–electrode interface is significantly reduced by the presence of hair. Maturing MEMS/NEMS technologies also hold promise as novel actuation devices and state-measurement devices.

EEG acquisition devices are usually designed for low power consumption and low cost. The acquired signals are transmitted to another device with a high-end processor through a transmission medium. Currently, most projects use conventional wireless technology such as Bluetooth [27,81,82], 802.15.4/Zigbees [66,83] or RF transmitters [63,84]. Bluetooth was developed as a replacement for cables and is a common feature in laptops and cellular phones. However, it also has a heavy protocol stack and high power consumption. In contrast, the protocol stacks of RF transmitters and 820.15.4/Zigbees are light and more efficient in terms of their power use, but they are less common in laptops. Multichannel EEG transmission should have a high data rate and high power consumption. Compression technology can be used to reduce the data rate. Lossless compression schemes are reliable but have lower compression ratios [85]. Lossy compression schemes demonstrate compression ratios but have artifact issues [86].

Advances in software and signal processing techniques are also needed to make wearable BCI a reality. Inherent changes that appear in brain signals when transferring from calibration to feedback sessions are a challenging but critical issue in BCI applications. The feasibility and importance of the adaptation of feature extraction in a self-paced BCI paradigm has been attempted in [87]. A new supervised method has been developed that learns from a feedback session in order to construct a more appropriate feature space on the basis of the maximum mutual information principle between feedback signal, target signal, and EEG. A progressive adaptive model-based approach that yields significantly improved MI detection has also been recently developed by Zhang et al. [46,47]. Furthermore, zero training processing techniques have been developed for P300-based BCI in [88].

3.5 The future of wearable brain–computer interface

The evolution of civilization points to a progressive increase in the interrelations between human minds, where by "mind" we mean a set of processes carried out by the brain [89]. Communication between minds or brains of different individuals has been supported and constrained by the sensory and motor arsenals of our bodies. There is the possibility of a new era in which brains will dialogue in a more direct way [90]. Previous attempts to realize this vision include demonstrations of bidirectional computer–brain communication [91–93] and cortical-spinal communication [94] in the monkey, and hippocampus-to-hippocampus [95] or social communication [96] in the rat—all of invasive nature. Despite these and other significant advances with human subjects [97,98], invasive methods in humans remain severely limited in their practical usefulness. Recent work has demonstrated fully noninvasive human to rat B2B communication by combining motor imagery driven EEG in humans on the BCI side with ultrasound brain stimulation on the CBI-rat side [99]. However, those traditional sensors have the disadvantage of being uncomfortable and require conductive gel and skin preparation on the part of the user. Therefore, acquiring EEG signals in a comfortable and convenient manner is an important factor that should be incorporated into a novel BCI device. In the foreseeable future, wearable, wireless, and portable EEG-based BCI devices with dry foam-based EEG sensors will definitely be developed first for gaming applications. The dry EEG sensors will operate without conductive gel, will provide good conductivity, and will be able to acquire EEG signals effectively by adapting to irregular skin surfaces and by maintaining proper skin-sensor impedance on the scalp. Real-time cognitive stage detection applications for gaming control will also be extended to clinical and other day-to-day activities. Obeid et al. [72] employed a wireless local area network to link the front-end data acquisition circuit and the data analysis system in a multichannel neural telemetry system capable of continuously transmitting 12 of 16 channels. A miniature 486 PC equipped with an IEEE802.11b wireless Ethernet card was used to transmit digitized signals. Digitized data (up to 12 bits of resolution at 31,250 samples/s per channel) were transferred to the PC and sent to a nearby host computer on a wireless local area network. Matthews et al. [67] developed a wireless multichannel system for EEG measurements based on a hybrid (capacitive/resistive) bioelectrode technology. This system featured a miniature, ultralow power microprocessor-controlled data acquisition system and a miniaturized wireless transceiver that can operate for over 72 hours from 2 AAA batteries. The system acted as a wearable and wireless EEG recording system rather than a complete BCI system since it did not incorporate any specific data processing capabilities or applications [67]. In future developments BCI systems capable of high fidelity data acquisition through wireless telemetry and online signal processing will be common. These systems are expected to provide further insight into the complex brain functions of users performing ordinary tasks in natural body positions in real operational environments using WSNs.

Security of BCI systems has started receiving attention recently [100,101]. Security of BCI systems has relied mainly on traditional password-based user authentication [102]. Cusack et al. structured a man in the middle attack where fake packets were substituted for the real ones in a BCI system, revealing the lack of security in current BCI systems [103]. Information compromised by lapses of security in BCIs includes external control capability and can affect the feedback loop to the user. If either of these two information streams is compromised, current BCI systems will experience unplanned for consequences arising from the use of the technology.

EEG-based authentication has been attempted recently [57,58,104]. Ruiz-Blondet et al. has attempted a novel method for very high accuracy event-related potential biometric identification through averaged event-related potentials in [104]. Del Pozo-Banos et al. have attempted subject identification through EEG [100]. Therefore, more research is needed into security and ethical issues related to BCI systems to make wearable BCI systems practical [105].

References

[1] E. Donchin, K.M. Spencer, R. Wijesinghe, The mental prosthesis: assessing the speed of a P300-based brain-computer interface, IEEE Transact. Rehab. Eng. 8 (2) (2000) 174–179.

[2] C. Grau, R. Ginhoux, A. Riera, T.L. Nguyen, H. Chauvat, M. Berg, et al., Conscious brain-to-brain communication in humans using non-invasive technologies, PLoS One 9 (8) (2014) e105225.

[3] W.G. Walter, Slow potential waves in the human brain associated with expectancy, attention and decision, Eur. Archiv. Psychiat. Clin. Neurosci. 206 (3) (1964) 309–322.

[4] J.J. Vidal, Towards direct brain–computer communication, Annu. Rev. Biophys. Bioeng. 2 (1973) 157–180.

[5] J.R. Wolpaw, The BCI endeavor and the mission of this new journal, Brain-Comput. Interf. 1 (1) (2014) 2–4.

[6] J.R. Wolpaw, N. Birbaumer, D.J. McFarland, G. Pfurtscheller, T.M. Vaughan, Brain–computer interfaces for communication and control, Clin. Neurophysiol. 113 (6) (2002) 767–791.

[7] L.A. Farwell, E. Donchin, Talking off the top of your head: toward a mental prosthesis utilizing event-related brain potentials, Electroencephal. Clin. Neurophysiol. 70 (6) (1988) 510–523.

[8] N. Bolognini, A. Pascual-Leone, F. Fregni, Using non-invasive brain stimulation to augment motor training-induced plasticity, J. Neuroeng. Rehabil. 6 (2009) 8.

[9] B.H. Dobkin, Training and exercise to drive poststroke recovery, Nat. Clin. Pract. Neurol. 4 (2008) 76–85.

[10] M.B. Buntin, Access to postacute rehabilitation, Arch. Phys. Med. Rehabil. 88 (2007) 1488–1493.

[11] R. Mane, E. Chew, K.S. Phua, K.K. Ang, A.P. Vinod, C. Guan, Quantitative EEG as biomarkers for the monitoring of post-stroke motor recovery in BCI and tDCS Rehabilitation, in: 2018 40th Annual International Conference of the IEEE Engineering in Medicine and Biology Society (EMBC), IEEE, 2018, pp. 3610–3613.

[12] E.W. Sellers, P. Turner, W.A. Sarnacki, T. McManus, T.M. Vaughan, R. Matthews, A novel dry electrode for brain-computer interface, Lect. Notes Comput. Sci. 5611 (2009) 623–631.

[13] S. Sutton, M. Braren, J. Zubin, E.R. John, Evoked-potential correlates of stimulus uncertainty, Science 150 (3700) (1965) 1187–1188.

[14] S. Soderlund, Interactive Productline AB, Method for playing games using brain waves, US Patent 7,988,557, 2011.

[15] R. Lievesley, M. Wozencroft, D. Ewins, The Emotiv EPOC neuroheadset: an inexpensive method of controlling assistive technologies using facial expressions and thoughts? J. Assist. Technol. 5 (2) (2011) 67–82.

[16] J. Katona, I. Farkas, T. Ujbanyi, P. Dukan, A. Kovari. Evaluation of the NeuroSky MindFlex EEG headset brain waves data, in: 2014 IEEE 12th International Symposium on Applied Machine Intelligence and Informatics (SAMI), IEEE, 2014, pp. 91–94.

[17] J.S. Ebersole, T.A. Pedley, Current Practice of Clinical Electroencephalography, third ed., Lippincott Williams & Wilkins, 2003, pp. 72–99.

[18] T. Alotaiby, F.E.A. El-Samie, S.A. Alshebeili, I. Ahmad, A review of channel selection algorithms for EEG signal processing, EURASIP J. Adv. Signal Process. 2015 (1) (2015) 66.

[19] J. Van Erp, F. Lotte, M. Tangermann, Brain-computer interfaces: beyond medical applications, Computer 45 (4) (2012) 26–34.

[20] D.R. Beukelman, P. Mirenda, Augmentative & Alternative Communication, Paul H. Brookes Publishing Co., Inc., Baltimore, MD, 2013.

[21] B.S. Oken, U. Orhan, B. Roark, D. Erdogmus, A. Fowler, A. Mooney, et al., Brain–computer interface with language model–electroencephalography fusion for locked-in syndrome, Neurorehab. Neural Repair 28 (4) (2014) 387–394.

[22] R. Ortner, F. Aloise, R. Prückl, F. Schettini, V. Putz, J. Scharinger, et al., Accuracy of a P300 speller for people with motor impairments: a comparison, Clin. EEG Neurosci. 42 (2011) 214–218.

[23] M. Sengelmann, A.K. Engel, A. Maye, Maximizing information transfer in ssvEP-based brain–computer interfaces, IEEE Trans. Biomed. Eng. 64 (2) (2017) 381–394.

[24] E. Vaadia, N. Birbaumer, Grand challenges of brain computer interfaces in the years to come, Front. Neurosci. 3 (2009) 15.

[25] M. Rizk, C.A. Bossetti, T.A. Jochum, S.H. Callender, M.A. Nicolelis, D.A. Turner, et al., A fully implantable 96-channel neural data acquisition system, J. Neural Eng. 6 (2009) 026002.

[26] E. Balanou, M. van Gils, T. Vanhala, State-of-the-art of wearable EEG for personalized health applications, in: PHealth 2013: Proceedings of the 10th International Conference on Wearable Micro and Nano Technologies for Personalized Health, vol. 189, IOS Press, 2013, pp. 119–124.

[27] C.-T. Lin, L.-W. Ko, M.-H. Chang, J.-R. Duann, J.-Y. Chen, T.-P. Su, et al., Review of wireless and wearable electroencephalogram systems and brain-computer interfaces—a mini-review, Gerontology 56 (1) (2009) 112–119.

[28] A. Nag, S.C. Mukhopadhyay, J. Kosel, Wearable flexible sensors: a review, IEEE Sens. J. 17 (13) (2017) 3949–3960.

[29] A. Dodabalapur, Welcome to flexible and printed electronics, Flexible Print. Electron. 1 (1) (2016) 010201.

[30] E. Cantatore, Applications of Organic and Printed Electronics, Springer, New York, 2013.

[31] O. Bubnova, Printed electronics: nanotube resolution, Nat. Nanotechnol. (2017).

[32] J.S. Chang, A.F. Facchetti, R. Reuss, Guest editorial organic/printed electronics: a circuits and systems perspective, IEEE J. Emerg. Selected Topic. Circuit. Syst. 7 (1) (2017) 1–6.

[33] J.A. Lewis, B.Y. Ahn, Device fabrication: three-dimensional printed electronics, Nature 518 (7537) (2015) 42–43.

[34] M. Stoppa, A. Chiolerio, Wearable electronics and smart textiles: a critical review, Sensors 14 (7) (2014) 11957–11992.

[35] C.T. Lin, C.J. Chang, B.S. Lin, S.H. Hung, C.F. Chao, I.J. Wang, A real-time wireless brain–computer interface system for drowsiness detection, IEEE Trans. Biomed. Circuit. Syst. 4 (4) (2010) 214–222.

[36] J.J. Norton, D.S. Lee, J.W. Lee, W. Lee, O. Kwon, P. Won, et al., Soft, curved electrode systems capable of integration on the auricle as a persistent brain–computer interface, Proceedings of the National Academy of Sciences, 2015, . 3920–3925.

[37] X. Huo, M. Ghovanloo, Evaluation of a wireless wearable tongue–computer interface by individuals with high-level spinal cord injuries, J. Neural Eng. 7 (2) (2010) 026008.

[38] L. Mayaud, M. Congedo, A. Van Laghenhove, D. Orlikowski, M. Figere, E. Azabou, et al., A comparison of recording modalities of P300 event-related potentials (ERP) for brain-computer interface (BCI) paradigm, Neurophys. Clin. 43 (2013) 217–227.

[39] C. Guger, G. Krausz, B.Z. Allison, G. Edlinger, Comparison of dry and gel based electrodes for P300 brain–computer interfaces, Front. Neurosci. 6 (2012) 60.

[40] G. Schalk, K.J. Miller, N.R. Anderson, J.A. Wilson, M.D. Smyth, J.G. Ojemann, et al., Two dimensional movement control using electrocorticographic signals in humans, J. Neural Eng. 5 (2008) 75–84.

[41] M.L. Homer, A.V. Nurmikko, J.P. Donoghue, L.R. Hochberg, Sensors and decoding for intracortical brain computer interfaces, Annu. Rev. Biomed. Eng. 15 (2013) 383–405.

[42] J.E. Huggins, C. Guger, B. Allison, C.W. Anderson, A. Batista, A.M. Brouwer, et al., Workshops of the fifth international brain–computer interface meeting: defining the future, Brain Comput. Interface 1 (2014) 27–49.

[43] T. Sekitani, S. Yoshimoto, T. Araki, T. Uemura, 12-2: Invited paper: a sheet-type wireless electroencephalogram (EEG) sensor system using flexible and stretchable electronics, in: SID Symposium Digest of Technical Papers, vol. 48, no. 1, 2017, May, pp. 143–146.

[44] K.B. Mikkelsen, P. Kidmose, L.K. Hansen, On the keyhole hypothesis: high mutual information between ear and scalp EEG, Front. Human Neurosci. 11 (2017) 341.

[45] C.T. Lin, C.Y. Chiu, A.K. Singh, J.T. King, Y.K. Wang, A wireless multifunctional SSVEP-based brain computer interface assistive system, IEEE Trans. Cognit. Develop. Syst (Early access).

[46] Q. Zhang, P. Wang, Y. Liu, B. Peng, Y. Zhou, Z. Zhou, et al., A real-time wireless wearable electroencephalography system based on support vector machine for encephalopathy daily monitoring, Int. J. Distrib. Sens. Netw. 14 (5) (2018). 1550147718779562.

[47] Z. Zhang, R. Foong, K.S. Phua, C. Wang, K.K. Ang, Modeling EEG-based motor imagery with session to session online adaptation, in: 2018 40th Annual International Conference of the IEEE Engineering in Medicine and Biology Society (EMBC), IEEE, 2018, pp. 1988–1991.

[48] C.Y. Chiu, A.K. Singh, Y.K. Wang, J.T. King, C.T. Lin, A wireless steady state visually evoked potential-based BCI eating assistive system, in: 2017 International Joint Conference on Neural Networks (IJCNN), IEEE, 2017, May, pp. 3003–3007.

[49] J. Liu, L. Yi, Liquid metal enabled wearable electronics, Liquid Metal Biomaterials, Springer, Singapore, 2018, pp. 369–416.

[50] H. Gui, J. Liu, Latest progresses in developing wearable monitoring and therapy systems for managing chronic diseases. arXiv preprint arXiv:1802.01747, 2018.

[51] H.C. Koydemir, A. Ozcan, Wearable and implantable sensors for biomedical applications, Annu. Rev. Anal. Chem. 11 (2018). Available from: https://doi.org/10.1146/annurev-anchem-061417-125956.

[52] P. Kumari, L. Mathew, P. Syal, Increasing trend of wearables and multimodal interface for human activity monitoring: a review, Biosens. Bioelectron. 90 (2017) 298–307.

[53] O. Faust, Y. Hagiwara, T.J. Hong, O.S. Lih, U.R. Acharya, Deep learning for healthcare applications based on physiological signals: a review, Comput. Method. Progr. Biomed. 161 (2018) 1–13.
[54] K. Mohanavelu, R. Lamshe, S. Poonguzhali, K. Adalarasu, M. Jagannath, Assessment of human fatigue during physical performance using physiological signals: a review, Biomed. Pharmacol. J. 10 (4) (2017) 1887–1896.
[55] V. Thayananthan, A. Basuhail, Integration of wearable smart sensor for improving e-healthcare, Integration 8 (12) (2017).
[56] Y. Ma, Y. Wang, J. Yang, Y. Miao, W. Li, Big health application system based on health Internet of things and big data, IEEE Access, 2016.
[57] M. Chen, Y. Ma, Y. Li, Y. Zhang, C.-H. Youn, Wearable 2.0: Enable human-cloud integration in next generation healthcare system, IEEE Commun. Mag. 54 (2016) 11.
[58] Y. Chen, A.D. Atnafu, I. Schlattner, W.T. Weldtsadik, M.C. Roh, H.J. Kim, et al., A high-security EEG-based login system with RSVP stimuli and dry electrodes, IEEE Trans. Inform. Foren. Security 11 (12) (2016) 2635–2647.
[59] E. Aguirre, P. Lopez-Iturri, L. Azpilicueta, C. Rivarés, J.J. Astrain, J. Villadangos, et al., Design and performance analysis of wireless body area networks in complex indoor e-Health hospital environments for remote patient monitoring, Int. J. Distrib. Sens. Netw. 12 (9) (2016). p. 1550147716668063.
[60] A. Casson, D. Yates, S. Smith, J. Duncan, E. Rodriguez-Villegas, Wearable electroencephalography, IEEE Eng. Med. Biol. Mag. 29 (2010) 44–56.
[61] Y.M. Chi, P. Ng, E. Kang, J. Kang, J. Fang, G. Cauwenberghs, Wireless non-contact cardiac and neural monitoring, in: Proceedings of the Wireless Health 2010, San Diego, CA, 2010, pp. 15–23.
[62] Y.M. Chi, T.P. Jung, G. Cauwenberghs, Dry-contact and noncontact biopotential electrodes: methodological review, IEEE Rev. Biomed. Eng. 3 (2010) 106–119.
[63] T. Torfs, V. Leonov, R.F. Yazicioglu, P. Merken, C.V. Hoof, R.J.M. Vullers, et al., Wearable autonomous wireless electro-encephalography system fully powered by human body heat, in: Proceedings of the 2008 IEEE Sensors, 2008, pp. 1269–1272.
[64] Y. Liu, M. Pharr, G.A. Salvatore, Lab-on-skin: a review of flexible and stretchable electronics for wearable health monitoring, ACS Nano 11 (10) (2017) 9614–9635.
[65] M. Abo-Zahhad, S.M. Ahmed, S.N. Abbas, State-of-the-art methods and future perspectives for personal recognition based on electroencephalogram signals, IET Biometr. 4 (3) (2015) 179–190.
[66] A. Riera, S. Dunne, I. Cester, G. Ruffini, BSTARFAST: a wireless wearable EEG/ECG biometric system based on the ENOBIO sensor, in: Proceedings of the 5th Health Workshop Wearable Micro Nanosyst. Pers. Health (pHealth'08), 2008.
[67] R. Matthews, N. McDonald, H. Anumula, J. Woodward, P. Turner, M. Steindorf, et al., Novel hybrid bioelectrodes for ambulatory zero-prep EEG measurements using multi-channel wireless EEG system, Lect. Notes Comput. Sci. 4565 (2007) 137–146.
[68] M.K. Mukerjee, NeuroPhone: brain-mobile phone interface using a wireless EEG headset, MobiHeld, 2010.
[69] L. Brown, J.V.D. Molengraft, R.F. Yazicioglu, T. Torfs, J. Penders, and C.V. Hoof, BA low-power, wireless, 8-channel EEG monitoring headset, in: Proceedings of the IEEE 2010 Annual International Conference on IEEE Engineering in Medicine and Biology Society (EMBC), 2010, pp. 4197–4200.

[70] W.D. Hairston, K.W. Whitaker, A.J. Ries, J.M. Vettel, J.C. Bradford, S.E. Kerick, et al., Usability of four commercially oriented EEG systems, J. Neural Eng. 11 (2014) 046018.

[71] A.K. Whitchurch, B.H. Ashok, R.V. Kumaar, K. Sarukesi, V.K. Varadan, Wireless system for long-term EEG monitoring of absence epilepsy, Biomed. Appl. Micro. Nanoeng. 4937 (2002) 343–349.

[72] I. Obeid, M.A.L. Nicolelis, P.D. Wolf, A multichannel telemetry system for signal unit neural recordings, J. Neurosci. Methods 133 (2004) 33–38.

[73] F. Cincotti, L. Kauhanen, F. Aloise, T. Palomaki, N. Caporusso, P. Jylanki, et al., Vibrotactile feedback for brain-computer interface operation, Comput. Intell. Neurosci. (2007) 48937.

[74] A.M. Chiarelli, F. Zappasodi, F. Di Pompeo, A. Merla, Simultaneous functional near-infrared spectroscopy and electroencephalography for monitoring of human brain activity and oxygenation: a review, Neurophotonics 4 (4) (2017) 041411.

[75] G. Ruffini, S. Dunne, E. Farrés, J. Marco-Pallarés, C. Ray, E. Mendoza, et al., A dry electrophysiology electrode using CNT arrays, Sens. Actuat. A: Phys. 132 (2006) 34–41.

[76] G. Anna, S. Hansen, J. Müller, Novel dry electrodes for ECG monitoring, Physiologic. Measur. 28 (2007) 1375.

[77] J.-Y. Baek, J.-H. An, J.-M. Choi, K.-S. Park, S.-H. Lee, Flexible polymeric dry electrodes for the long-term monitoring of ECG, Sens. Actuat. A Phys. 143 (2008) 423–429.

[78] T.I. Oh, S. Yoon, T.E. Kim, H. Wi, K.J. Kim, E.J. Woo, et al., Nanofiber web textile dry electrodes for long-term biopotential recording, IEEE Trans. Biomed. Circ. Syst. 7 (2) (2013) 204–211.

[79] A.C. Myers, H. Huang, Y. Zhu, Wearable silver nanowire dry electrodes for electrophysiological sensing, RSC Adv. 5 (15) (2015) 11627–11632.

[80] G. Li, D. Zhang, S. Wang, Y.Y. Duan, Novel passive ceramic based semi-dry electrodes for recording electroencephalography signals from the hairy scalp, Sens. Actuat. B Chem. 237 (2016) 167–178.

[81] Y.T. Wang, Y. Wang, T.P. Jung, A cell-phone-based brain–computer interface for communication in daily life, J. Neural Eng. 8 (2) (2011) 025018.

[82] B. Kerous, F. Skola, F. Liarokapis, EEG-based BCI and video games: a progress report, Virtual Real. 22 (2) (2018) 119–135.

[83] S. Narayanan, K.V. Divya, A sensor based mechanism for controlling mobile robots with ZigBee, Computational Intelligence in Data Mining—Volume 2, Springer, New Delhi, 2016, pp. 1–12.

[84] P.V. Patel, M. Sarkar, S. Nagaraj, K. Kushalad, Channel modelling based on statistical analysis for brain-computer-interface (BCI) applications, in: 2016 IEEE Conference on Computer Communications Workshops (INFOCOM WKSHPS), IEEE, 2016, pp. 320–321.

[85] M. Elgendi, A. Mohamed, R. Ward, Efficient ECG compression and QRS detection for e-Health applications, Sci. Rep. 7 (1) (2017) 459.

[86] P. Worgan, L. Clare, P. Proynov, B.H. Stark, D. Coyle, Inductive power transfer for on-body sensors, in: 10th EAI International Conference on Pervasive Computing Technologies for Healthcare, 2016, pp. 16–19.

[87] H. Zhang, S.R. Liyanage, C. Wang, C. Guan, Learning from feedback training data at a self-paced brain–computer interface, J. Neural Eng. 8 (4) (2011) 046035.
[88] A.M. Eldeib, O. Sarhan, M.A. Wahed, Zero training processing technique for P300-based brain-computer interface, in: 2018 IEEE 4th Middle East Conference on Biomedical Engineering (MECBME), IEEE, 2018, pp. 112–115.
[89] E. Kandel, The new science of mind and the future of knowledge, Neuron 80 (2013) 546–560.
[90] M.A. Nicolelis, Beyond Boundaries: The New Neuroscience of Connecting Brains With Machines and How It Will Change Our Lives, St. Martin's Griffin, 2010. 368 pp.
[91] J.K. Chapin, K.A. Moxon, R.S. Markowitz, M.A. Nicolelis, Real-time control of a robot arm using simultaneously recorded neurons in the motor cortex, Nat. Neurosci. 2 (1999) 664–670.
[92] J.M.R. Delgado, The Physical Control of the Mind: Toward a Psychocivilized Society, Harper & Row, New York, 1969. 288 pp.
[93] J.E. O'Doherty, M. Lebedev, P.J. Ifft, K.Z. Zhuang, S. Shokur, et al., Active tactile exploration using a brain–machine–brain interface, Nature 479 (2011) 228–231.
[94] M.M. Shanechi, R.C. Hu, Z.M. Williams, A cortical-spinal prosthesis for targeted limb movement in paralysed primate avatars, Nat. Commun. 5 (2014) 3237.
[95] S.A. Deadwyler, T.W. Berger, A.J. Sweatt, D. Song, R.H. Chan, et al., Donor/recipient enhancement of memory in rat hippocampus, Front. Syst. Neurosci. 7 (2013) 120.
[96] M. Pais-Vieira, M. Lebedev, C. Kunicki, J. Wang, M.A.L. Nicolelis, Brain-to-brain interface for real-time sharing of sensorimotor information, Sci. Rep. 3 (2013) 1319.
[97] L.R. Hochberg, D. Bacher, B. Jarosiewicz, N.Y. Masse, J.D. Simeral, et al., Reach and grasp by people with tetraplegia using a neurally controlled robotic arm, Nature 485 (2012) 372–375.
[98] L.R. Hochberg, M.D. Serruya, G.M. Friehs, J.A. Mukand, M. Saleh, et al., Neuronal ensemble control of prosthetic devices by a human with tetraplegia, Nature 442 (2006) 164–171.
[99] S.-S. Yoo, H. Kim, E. Filandrianos, S.J. Taghados, S. Park, Non-invasive brain-to-brain interface (BBI): establishing functional links between two brains, PLoS One 8 (4) (2013) e60410.
[100] M. Del Pozo-Banos, J.B. Alonso, J.R. Ticay-Rivas, C.M. Travieso, Electroencephalogram subject identification: a review, Expert Syst. Appl. 41 (15) (2014) 6537–6554.
[101] Q. Li, D. Ding, M. Conti, Brain-computer interface applications: security and privacy challenges, in: 2015 IEEE Conference on Communications and Network Security (CNS), IEEE, 2015, pp. 663–666.
[102] Q. Li, Human-computer interaction: security aspects, 2018.
[103] B. Cusack, K. Sundararajan, R. Khaleghparast, Neurosecurity for brainware devices, 2017.
[104] M.V. Ruiz-Blondet, Z. Jin, S. Laszlo, CEREBRE: a novel method for very high accuracy event-related potential biometric identification, IEEE Trans. Inform. Foren. Security 11 (7) (2016) 1618–1629.
[105] M. Ienca, P. Haselager, Hacking the brain: brain–computer interfacing technology and the ethics of neurosecurity, Ethics Inform. Technol. 18 (2) (2016) 117–129.

Further reading

D.S.V. Bandara, J. Arata, K. Kiguchi, A noninvasive brain–computer interface approach for predicting motion intention of activities of daily living tasks for an upper-limb wearable robot, Int. J. Adv. Rob. Syst. 15 (2) (2018). p. 1729881418767310.

L. Beckmann, C. Neuhaus, G. Medrano, N. Jungbecker, M. Walter, T. Gries, et al., Characterization of textile electrodes and conductors using standardized measurement setups, Physiol. Measur. 31 (2010) 233.

B. Blankertz, M. Tangermann, C. Vidaurre, S. Fazli, C. Sannelli, S. Haufe, et al., The Berlin brain–computer interface: non-medical uses of BCI technology, Front. Neurosci. 4 (2010) 198.

M. Cheng, X. Gao, S. Gao, D. Xu, Design and implementation of a brain-computer interface with high transfer rates, IEEE Trans. Biomed. Eng. 49 (2002) 1181–1186.

O. Dehzangi, M. Farooq, Wearable brain computer interface (BCI) to assist communication in the intensive care unit (ICU), in: 2018 IEEE International Conference on Consumer Electronics (ICCE), IEEE, 2018, pp. 1–4.

A. Fleury, M. Alizadeh, G. Stefan, T. Chau, Toward fabric-based EEG access technologies: seamless knit electrodes for a portable brain-computer interface, in: Life Sciences Conference (LSC), 2017 IEEE, IEEE, 2017, pp. 35–38.

S. Gao, Y. Wang, X. Gao, B. Hong, Visual and auditory brain–computer interfaces, IEEE Trans. Biomed. Eng. 61 (5) (2014) 1436–1447.

B. Graimann, B.Z. Allison, G. Pfurtscheller (Eds.), Brain-Computer Interfaces: Revolutionizing Human-Computer Interaction, Springer Science & Business Media, 2010.

P. Griss, P. Enoksson, H.K. Tolvanen-Laakso, P. Meriläinen, S. Ollmar, G. Stemme, BMicromachined electrodes for biopotential measurement, IEEE J. Microelectromech. Syst. 10 (2001) 10–15.

M.U. Hassan, J. Keck, H. Klauk, J. Kostelnik, Y. Mahsereci, S. Sailer, A. Schreivogel, T. Zaki, J.N. Burghartz, Combining organic and printed electronics in Hybrid System in Foil (HySiF) based smart skin for robotic applications, 2015 European Microelectronics Packaging Conference (EMPC, IEEE, 2015, September, pp. 1–6.

L.D. Liao, C.T. Lin, K. McDowell, A.E. Wickenden, K. Gramann, T.P. Jung, et al., Biosensor technologies for augmented brain-computer interfaces in the next decades, Proc. IEEE 100 (2012) 1553–1566.

C.T. Lin, L.W. Ko, J.C. Chiou, J.R. Duann, R.S. Huang, S.F. Liang, et al., Noninvasive neural prostheses using mobile and wireless EEG, Proc. IEEE 96 (2008) 7.

R. Matthews, P.J. Turner, N.J. McDonald, K. Ermolaev, T.M. Manus, R.A. Shelby, et al., Real time workload classification from an ambulatory wireless EEG system using hybrid EEG electrodes, in: Proceedings of the IEEE 30th Annual International Conference, on Engineering in Medicine and Biology Society, EMBC 2008, 2008, pp. 5871–5875.

F. Nijboer, B. Van De Laar, S. Gerritsen, A. Nijholt, M. Poel, Usability of three electroencephalogram headsets for brain–compute interfaces: a within subject comparison, Interact. Comput. 27 (2015) 500–511.

S. Stirenko, G. Peng, W. Zeng, Y. Gordienko, O. Alienin, O. Rokovyi, et al., Parallel statistical and machine learning methods for estimation of physical load. arXiv preprint arXiv:1808.04760, 2018.

K. Suganuma, Introduction to Printed Electronics, Springer Science & Business Media, 2014.

D. Tan, A. Nijholt, Brain-computer interfaces and human-computer interaction, Brain-Computer Interfaces, Springer, London, 2010, pp. 3–19.

J.F. Valenzuela-Valdes, M.A. Lopez, P. Padilla, J.L. Padilla, J. Minguillon, Human neuro-activity for securing body area networks: application of brain-computer interfaces to people-centric Internet of things, IEEE Commun. Mag. 55 (2) (2017) 62–67.

D. Wu, B.J. Lance, V.J. Lawhern, Guest editorial for the special section on brain computer interface (BCI), IEEE Trans. Fuzzy Syst. 25 (1) (2017) 1–2.

C. Yeom, K. Chen, D. Kiriya, Large-area compliant tactile sensors using printed carbon nanotube active-matrix backplanes, Adv. Mater. 27 (9) (2015) 1561-–1566.

S. Zhang, S. Li, S. Cheng, Research on smart sensing RFID tags under flexible substrates in printed electronics, pp. 1006–1009.

CHAPTER

AdaptableSDA: secure data aggregation framework in wireless body area networks

4

Vivaksha J. Jariwala[1] and Devesh C. Jinwala[2]

[1]*Sarvajanik College of Engineering and Technology, Surat, India*

[2]*S.V. National Institute of Technology, Surat, India*

Chapter Outline

4.1 Introduction

A wireless body area network (WBAN) is composed of small, intelligent devices that may be implanted or mounted on the human body [1–3]. These devices transmit data to a central patient monitoring station, known as a base

Wearable and Implantable Medical Devices. DOI: https://doi.org/10.1016/B978-0-12-815369-7.00004-5

station (BS), where the data is analyzed to determine the appropriate course of action.

A WBAN is useful for various applications of wearable and implantable medical devices. A WBAN has two primary aspects: (1) data collection via implantable sensors [8] and (2) analysis of the collected data for secure, high-quality diagnosis [4–7]. Recent healthcare applications of WBANs include monitoring conditions such as asthma, cardiovascular disease, diabetes, Alzheimer's disease, depression, and epileptic seizures. WBANs are also utilized for cancer detection, stress tracking, monitoring of the elderly, medical accident prevention, and vision restoration (artificial retina) [9].

The devices in a WBAN typically have restricted memory and communication and computational proficiencies due to size and mobility requirements. Further, the need for continuous operation with limited energy replenishment requires minimal energy expenditure. In-network processing is a design paradigm used to overcome these challenges.

In-network processing depends on processing the sensed raw data in real time, as it is being communicated to the external node (EN)/BS [10,11]. In in-network processing, the intermediate SNs are required to process the data in the incoming packet before deciding whether to propagate the packet further or not. This type of multihop communication is also known as *data-centric* multihop communication [12]. Data-centric multihop communication is different from conventional *address-centric* multihop communication, wherein the intermediate nodes only inspect the address in the packet header to determine the further route of the packet. The advantage of using in-network processing is a reduction in the overall number of packets communicated to the BS as illustrated in Fig. 4.1.

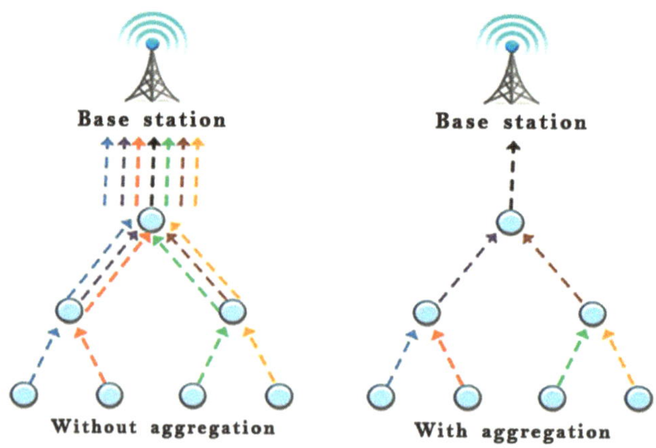

FIGURE 4.1

Wireless sensor networks (WSNs) with and without aggregation.

WBANs must satisfy security and privacy requirements, which depend on the type of application, either medical or nonmedical. In the case of medical applications, security breaches may result in critical harm to patients or even death. Hence, scalable and stringent security mechanisms are required to ensure security in WBANs.

In the case of in-network processing in WBANs, a consequence of exposing data for processing to the intermediate nodes in a network is that the security and privacy of the data may be compromised. Hence, it is necessary that when using in-network processing for data aggregation, appropriate security mechanisms are put in place. Such aggregation of data is referred to as secure data aggregation (SDA). SDA in WBANs should include confidentiality, privacy, integrity, authentication, and key establishment protocol.

In this chapter, we propose an adaptable integrated framework called the AdaptableSDA for SDA in WBANs (Fig. 4.2). AdaptableSDA has two modules. The first module provides support for the desired security attributes, i.e., privacy and confidentiality during the data aggregation operation. Since it is also necessary to ensure data integrity and entity authentication, the second module offers these two attributes.

As a digital signature provides nonrepudiation property, we consider digital signature as a mechanism for providing end-to-end integrity. We used the elliptic curve digital algorithm (ECDSA) as a candidate algorithm. The motivation for using ECDSA includes its smaller key size that provides the same level of security. This small key size makes ECDSA useful for implementation in hardware such as SNs. An additional advantage of ECDSA is the computational hardness of breaking the elliptic curve discrete logarithm problem (ECDLP), which is incorporated in the algorithm.

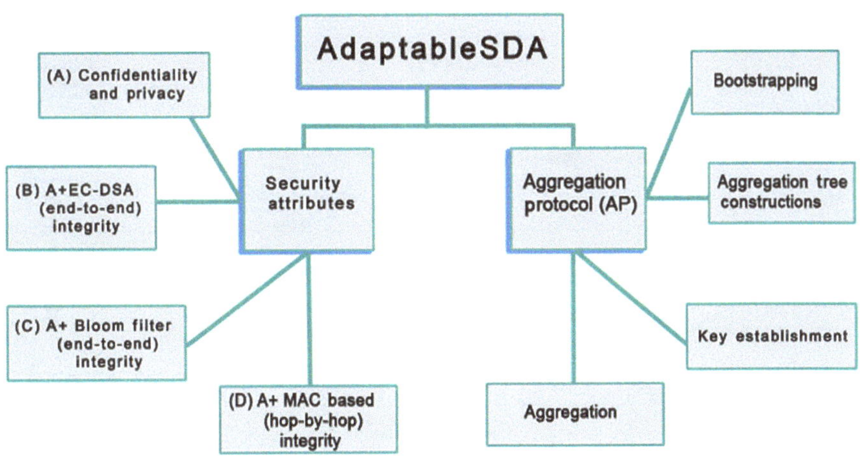

FIGURE 4.2

Framework—AdaptableSDA.

In order to enhance the security strength of the basic EC-DSA, we also propose two variants of ECDSA based on: (1) three random variables and (2) Bloom filters. Moreover, to decrease the overhead due to the digital signatures and to make AdaptableSDA more versatile, the authors propose an integrity check scheme that relies on the use of a message authentication code (MAC) for SDA in AdaptableSDA. The proposed scheme performs the integrity checks at every hop in the network, thus eliminating the possibility of any intermediate node wasting energy in communicating an invalid data value to its parent node.

Using theoretical analysis supplemented by experimentation on the TinyOS platform, we show that it is appropriate to employ this variant of EC-DSA in the resource-constrained environment of WSNs. We further substantiate the proposal with elaborate experimental evaluation.

4.2 Background

Privacy homomorphism is a mechanism that allows direct computation of encrypted data [13]. This approach not only supports privacy, but also reduces vulnerabilities to attacks due to the infrequent cipher operations and thus enables reduction in overhead. In conventional encryption aggregation can be performed only on plain data (Fig. 4.3). As compared, when using homomorphic encryption, aggregation is performed on the encrypted data; the ciphertext is not required to be decrypted to plaintext to perform aggregation on the plaintext (Fig. 4.4).

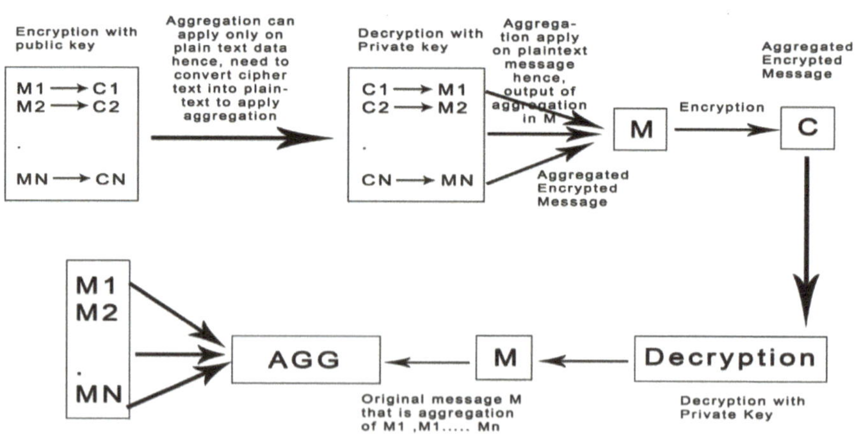

FIGURE 4.3

Conventional encryption where aggregation is performed on the plaintext data only.

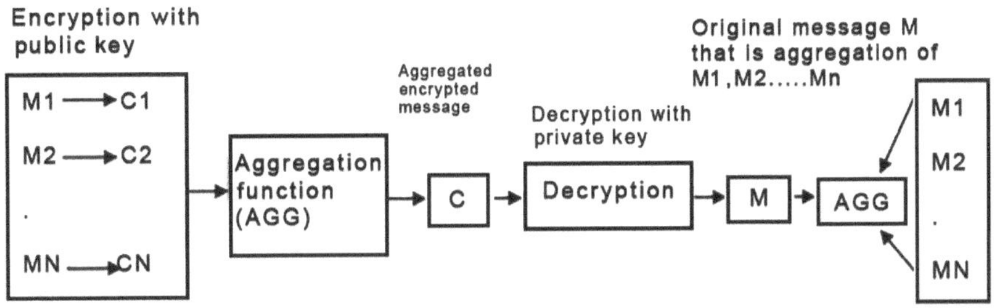

FIGURE 4.4

Homomorphic encryption where aggregation is performed directly on the encrypted data.

We categorize the published approaches for SDA as follows based on the security attributes and configurations they offer:

a. Category 1: Approaches that offer only confidentiality and privacy [14–29].
b. Category 2: Approaches that provide only integrity [30–42].
c. Category 3: Approaches that specifically do not focus on security attributes or key management for data aggregation [43–50].

The approaches in Category 1 are appropriate for numerous applications that require confidentiality and privacy, such as vehicular traffic analysis, location-tracking systems for vehicles or other devices, and healthcare monitoring systems.

Without data integrity protection mechanisms an application cannot ensure robustness of the data. All of the approaches mentioned are susceptible to data alteration attacks by malicious users or adversaries. Hence, it is clear that any security protocol without data integrity is not practical.

Though these approaches are notable attempts to provide confidentiality and privacy in SDA, their execution remains in question. As shown, the approaches in Category 2 provide only integrity, making them best suited for body area network and environmental monitoring applications. These applications may require monitoring of movement, blood pressure, temperature, or heartbeat (e.g., patient monitoring systems in hospitals). In such scenarios the most important security attribute is integrity of the aggregated. Such approaches cannot be used in applications demanding confidentiality of data, battlefield monitoring, or vehicular traffic analysis. Though these approaches offering only data integrity are notable attempts to achieve data integrity in a typical SDA protocol for body area networks, their concrete execution remains a concern.

Thus, the design of practical and implementable SDA system protocols for body area networks should offers confidentiality as well as integrity. There are approaches in the literature which provide integrity in addition to confidentiality [51,52]. The latter approach in [51] is a SDA scheme which uses elliptic curve cryptography (ECC) and offers integrity and confidentiality, but the scheme uses elliptic curve ElGamal (EC–EG) for encryption. The approaches discussed provide confidentiality, privacy, and integrity of the data and are therefore useful in patient monitoring systems. In patient monitoring, preservation of data confidentiality indicates an adversary cannot obtain the data while the data are being transferred. In the same scenario, preservation of data integrity indicates blocking of any attempts by malicious users to inject inaccurate data while data are being transferred. The approach proposed in [53] provides privacy as well as confidentiality, but lacks configurable options of required security attributes. Authors in [54] provide a survey of the approaches that provides privacy and integrity, but do not focus on configuration option. Authors in [55] also provide an approach for SDA with a focus on confidentiality and integrity, but not privacy.

Approaches specified in category 3 are diverse. The authors in [43] provide an approach for SDA with intrusion detection. The authors in [44] propose an SDA framework that provides support for confidentiality and integrity, but they do not focus on key management or topology of a body area network. In [45] the authors have proposed polynomial regression based SDA. Here the authors focus on reducing the amount of data transmission in the network without compromising data confidentiality. However, they do not focus on the integrity of the data. In [46] the author proposes SDA based on genetic algorithm. Authors in [47] propose an SDA scheme that preserves both integrity and confidentiality, but they do not empirically evaluate the proposed approach. The authors in [54] propose SDA based on fully homomorphic encryption and provide data integrity with the help of MAC. However, they do not provide configurable options for various applications. The authors in [48] propose SDA based on the use of interval analysis to detect and remove abnormal data. In [49] the authors propose a data-loss resilient data aggregation scheme that is based on key vector sharing. The authors in [50] focus on adaptive data aggregation. Their approach is based on a feedback control system that dynamically optimizes the tradeoff between the data aggregation process and network constraints. The authors in [56] provide an exhaustive survey of concealed data aggregation and discuss all the aspects of SDA, but do not focus on key management issues.

In our view all of the approaches referenced, remain largely proof-of-concept approaches—showing how specific security attributes can be achieved in a SDA setup. The focus of these previous publications is not on defining other peripheral aspects associated with a realistic SDA protocol, namely to define an appropriate topology and to define a key setup routine. Motivated by these lacunae, here we propose a complete security framework, AdaptableSDA, which offers a suite of security algorithms to choose from, uses a defined topology and allows selection of the desired security attributes.

The research attempts discussed in this chapter encompass the following:

1. We integrate EC-OU into a tree architecture and also provide end-to-end and hop-by-hop integrity with novel proposed variants of EC-DSA and MAC-based approaches, respectively.
2. We also provide topology construction and key distribution mechanisms to complete protocol.

4.2.1 Proposed work

In this section we outline the proposed work: the AdaptableSDA framework, which provides support for the subsequent features.

1. Provides support for privacy and confidentiality.
2. In addition to point (1) also provide support for end-to-end integrity based on EC-DSA.
3. In addition to point (1) also offer provision for end-to-end integrity based on a novel Bloom filter-based approach.
4. Support for point (1) and MAC based hop-by-hop integrity.

In our view a framework for an SDA must offer the minimum components of security attributes, topology construction, and key management. Without key management and topology construction, a framework for SDA cannot be complete. Prompted by this motivation, the second module of AdaptableSDA proposes an aggregation protocol (AP) that supports a topology construction protocol and an efficient key exchange protocol in addition to the desired security attributes.

The outline of the proposed framework is as follows:

1. It is assumed that the nodes form a tree topology of the network.
2. During formation of the tree topology, each node is expected to be aware of three levels in the hierarchy, that is, node a knows node c is its grandparent and node b is its parent.
3. After formation of the tree topology, the key setup process is initiated in which the keys for a node are derived using a transitory master key mechanism. That is each node shares key k_{parent} with its parent, $k_{grandparent}$ with its grandparent, and k_{child_i} with its immediate child nodes and two-hop child nodes.
4. When the aggregation phase starts, a parent node and a grandparent node receive identical copies of data from a specific child node, a. In order to ensure the confidentiality of a data item, x, it is sent as $Enc_{PubkBS}(x)$, where Enc is a privacy homomorphic encryption algorithm that uses $_{PubkBS}$, the public key of the BS.
5. In order to preserve integrity of the data item x, a message authentication code MAC(x) is sent along with the data item x in the packet communicated to the parent and the grandparent.
6. A grandparent node checks that the aggregation is valid by aggregating the data items received from its children and grandchildren and then equating the values received from children and the grandchildren. This guarantees that the grandparent node receives the same aggregation value as parent node and both values are compared.

4.3 Main focus of the chapter

Here we attempt to integrate basic EC-DSA and our own Bloom filter-based variant of EC-DSA to provide end-to-end integrity using the proposed AdaptableSDA with EC-OU for privacy. AdaptableSDA also provides support for a MAC based alternative to support hop-by-hop integrity. Hence, AdaptableSDA is more adaptable in that it supports EC-OU based privacy and confidentiality, MAC based option to provide hop-by-hop data integrity, and upgraded EC-DSA based signature to support end-to-end data integrity.

4.3.1 AdaptableSDA: end-to-end integrity

In this section we discuss integration of end-to-end integrity in AdaptableSDA. Support for end-to-end integrity is provided with two alternatives: (1) with EC-DSA and (2) with our own Bloom filter-based variant of ECDSA. ECDSA is used to provide end-to-end integrity in the AdaptableSDA. A 160-bit encryption key of ECC provides security for various applications and is equally secure compared to 1024-bit RSA key, yet is 15 times faster [57–60]. The aggregator node (AN) can aggregate sensed values from SNs and forward the final value to the BS. The data traveling from AN to BS must be secured during transmission. This approach of end-to-end integrity is useful in critical applications of health-related data transfer in hospitals where the BS is located in controlled environments away from the patient. In the patient monitoring system of body area networks, each SN generates signatures with the help of its private key and the BS can verify the signature with its own public key. In a body area network, SNs can detect changes in patient parameters such as temperature or blood pressure. If a change is detected, then flag (1) of that node can be transferred to the nearby AN where aggregation is applied. Finally, aggregation of changes can be forwarded to the BS. The information transferred by the SN is confidential information that should not be modified in transit and only the BS would able to verify the information received. Any intermediate node should not be able to verify the information received. Hence, in this case end-to-end integrity is essential. After receiving data from the AN, the BS can take appropriate actions. In this approach, hop-by-hop integrity is not useful, as an intermediate node does not have to take any actions after verifying integrity and if an intermediate node is captured by the enemy then information can be lost. A scenario where the approach of end-to-end integrity is essential, monitoring of changes in patient temperature or blood pressure, is shown in Fig. 4.5.

In the AdaptableSDA, SensorNodeAlgo() is deployed on sensor motes whereas the algorithm BaseStationAlgo() is deployed on the BS. In our variant of ECDSA, when the body area network is set up, the sender would create a Bloom filter data structure. Next, the sender generates a number called them as Bloom and it computes hash on the Bloom with the help of small hash functions [43]. Then the sender checks on the index of the hash value generated with the help of hash functions in the Bloom filter. If at that index, element is 1, it simply discards that and selects another variable Bloom. If on that index, the element is 0, then it uses the value of that Bloom. In this algorithm, each SN calculates the signature s_i on the outgoing message m_i.

$$s_i = \text{sign}(m_i) \tag{4.1}$$

After generating the signature, the SN also computes cipher-text c_i of the message.

$$c_i = \text{Enc}(m_i) \tag{4.2}$$

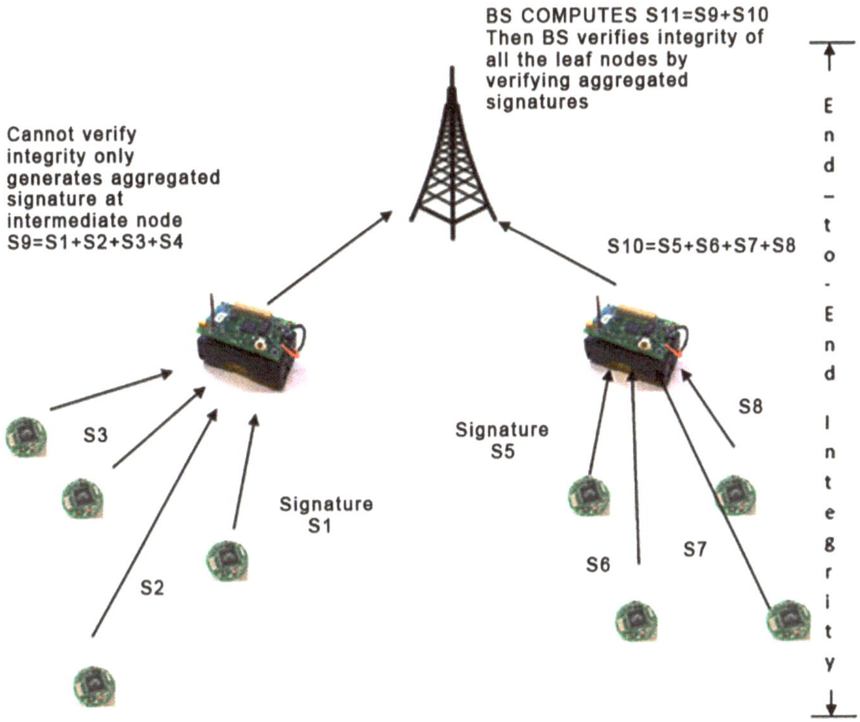

Figure 4.5 Example of a system where end-to-end integrity is essential.

If a SN is an AN, then it combines all the signatures received from the child nodes into one signature and all the cipher-texts received from the child node into one cipher-text.

Algorithm 1: Sensor_Node()

```
// Message m_i maps on the Elliptic_Curve E
// Parameters of Elliptic Curve E = (v, a, b, G, u, h)
// All sensor node calucalates the value
1.    Public Key  :  n = u^2 v, H, G, Q=uG
2.    Private Key :  u
3.    Signature Generation :
                        a. select variable bloom, where 1<=bloom<n
                        b. Create array of bloom_filter
                        c. Check existence of member in bloom_filter
                        d. If member in the bloom_filter go to step a
                           Else go to the next step e
                        e. BloomG=(x_1,y_1), r=x_1(mod n)
                        f. Each sensor computes signature
                          s_i=Bloom^-1(H(m)+ur)(mod n)
                           (r,s_i) is the signature
4.    Encryption  : plaintext m_i< 2^(k-1),r∈_R 2^(2k),
                   Cipher-text C_i = m_iG + rH
5.    if sensor node is anAggregator
        S = ∑ s_i // generate single signature by addition of all signature
       C = ∑ c_i // generate single cipher-text by addition of all cipher-texts
      end if
```

As in BaseStationAlgo(), BS receives aggregated cipher-texts and aggregated signatures from the SNs. The BS performs decryption of aggregated cipher-texts to obtain the plaintext message which is the sum of all plaintext messages. Likewise, the BS accepts the message only if the signature is verified. Hence, our approach provides privacy and confidentiality with the help of privacy homomorphic encryption. Our approach also ensures integrity through digital signature. Our approach provides privacy in that ANs only know that some movement is detected in a select region, but not by exactly which SN. Hence, if an adversary captures an AN, the adversary would still not be able to obtain the information. As discussed, the proposed approach provides end-to-end integrity in patient monitoring for body area networks. Though ECDSA is useful in providing integrity support for one-to-many traffic patterns [61], the result shows that it can also be efficiently used to provide integrity support for many-to-one traffic pattern of WSNs.

Algorithm 2: Base_Station()

```
// From the elliptic curve Emap cihper-text Ci
// Parameters of Elliptic Curve E = (v, a, b, G, u, h)
// The base station calculates the value

1. Public Key:n = u²v, G, Q=uG,H
2. Private Key: u
3. Calculate m =Ψp((u+2)C)/Ψp((u+2)G)(mod u)

          Ψp(x, y) = - x/y(mod u²)
4. BS get m = Σ mi
5. Compute w=s⁻¹(mod n)
          u1=H(m)w(mod n)
          u2=rw(mod n)
          u1G + u2Q = (x2,y2)
          v=x2(mod n)
6. If r==v then
          Signature is verified

 End if
```

4.3.2 AdaptableSDA: hop-by-hop integrity

MAC based hop-by-hop integrity for the AdaptableSDA is discussed in this section. In this approach, false data inserted by an adversary would be identified at the next hop and would not reach the BS. Whereas in the end-to-end approach, the false data would be identified at the destination only (i.e., at the BS), wasting energy and lifespan of the sensor networks. Wander et al. [62] investigated the energy cost for end-to-end data integrity verification with the help of an EC-DSA algorithm on an Atmel ATmega128 processor and found the EC-DSA algorithm requires 22.82 mJ to sign the message and 45.09 mJ to verify the message. In the earlier approach, signature verification is done only at the BS, which is rich in resources compared to a leaf node. Thus the authors' approach of EC-DSA is suitable for applications where end-to-end security is a necessity (patient monitoring). Our results show that it is feasible to deliver end-to-end integrity in body area network. In the proposed approach hop-by-hop integrity verification with MAC is discussed as a means to lower the resource overhead and increase the lifetime of body area network.

In AdaptableSDA, Sensor_Node() is deployed on sensor motes in the room where the monitoring is occurring, whereas the algorithm Base_Station() is deployed on the BS that would monitor the data received from that room. In this algorithm each SN calculates signature s_i on the outgoing message m_i.

$$c_i = \mathrm{Enc}(m_i) \tag{4.3}$$

SN generates cipher-text C_i on sensed message m_i.

Then, on generated C_i, MAC M_i is calculated.

$$M_i = \mathrm{MAC}(C_i) \tag{4.4}$$

The SN's parent receives message M_i and cipher-text C_i. The parent then generates message M_i from the received C_i. If computed C_i is the same as C_i received in the message, parent node accepts the message and performs aggregation on the messages. Once aggregation in complete, parent computes MAC on the aggregated data and sends to its parent. When BS receives a message, it computes MAC on the aggregated message and verifies whether computed MAC is the same as received MAC or not. If both the MACs are same, BS performs the decryption of aggregated cipher-text and generates the original message that is the summation of the all plaintext messages. Hence, the proposed approach provides privacy and confidentiality with the help of privacy homomorphic encryption and also provides data integrity.

Algorithm 3: Sensor_Node()

```
// Map plain-text message mi on the elliptic curve E
// Parameters of Elliptic Curve E = (v, a, b, G, u, h)
// The sensor node calucalates the value

Public Key   :  n = u²v, G,Q=uG, H
Private Key  :  u
Encryption   :  plaintext mi< 2^(k-1),r∈R 2^(2k),
                Cipher-text Ci = miG + rH
MAC Generation : The sensor node calculates MACi= MAC(Ci)
                if sensor node is the Aggregator
                  Calcuates MACi= MAC(Ci)
                    Verify Computed MACi  = received MACi
                    if above condition is satisfied
                        C = ∑ Ci //summation of ciphertexts into one ciphertext
                        and transfer to the parent node
                    end if
                end if
```

Algorithm 4: Base_Station()

```
// Map cihper-text Cifrom the elliptic curve E
// Parameters of Elliptic Curve E = (v, a, b, G, u, h)
// The base station calucalates the value

1. Public Key  :n = u²v, G, Q=uG , H
2. Private Key : u
3. MAC Verification: Calculate MACi= MAC(Ci)
                     Validate received MACi= Calculated MACi
                       if above condition is true go to next step
3. Calculate m =Ψp((p+2)C)/Ψp((p+2)G)(mod p)(mod p)

            Ψp(x, y) = - x/y(mod p²)
4. Base station would have m = ∑ mi
```

4.3.3 Proposed aggregation protocol

Table 4.1 shows symbolic notations used for the protocol.

The proposed aggregation protocol provides privacy with end-to-end confidentiality. In this approach privacy and confidentiality are preserved by the homomorphic encryption algorithm EC-OU. Each node sends its data encrypted by base station public key Key_{PU}. After receiving aggregated data from all child

Table 4.1 Notations of aggregation protocol.

Notation	Description
BSN	Base station
A_{ID}	Identification of sensor node A
A_{N1}, A_{N2}	Nonce generated by node A—A_{N1}, A_{N2}
OWF	One-way function—given two input seeds it calculates values of the same length as the key
$MasterK_A$	Master key of sensor node A
Key_{PR}	Base station private key
Key_{PU}	Base station public key
Key_{IN}	Initial key stored in sensor nodes
Mac(K, M)	MAC generated on message M with key K
Enc(K, M)	Message M encrypted with key K
$KMAC_{A,B}$	Key used for MAC generation and common to node A and B
AF	Aggregation function A
H	Message header
A→*	Broadcast message from node A
A_P	Parent of A
A_{GP}	Grandparent of A

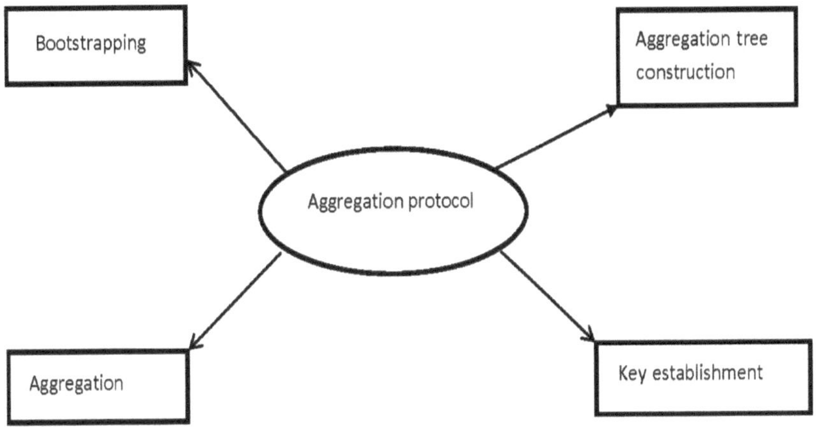

FIGURE 4.6
Overview of protocol.

nodes, the BS decrypts the aggregated data with its own private key Key_{PR}. In this approach data integrity is preserved by comparing calculated redundant aggregate values computed by two different types of sensor nodes. AdaptableSDA also provides authentication by means of sharing a secret key between the nodes which is derived from the transitory master key. The aggregation protocol has four parts: bootstrapping, tree construction for aggregation, establishment of keys, and aggregation. The first three parts of the aggregation protocol, that is, bootstrapping, tree construction and establishment of keys are executed only once for set-up. After completing the set-up phase, multiple time aggregation is applied to the constructed tree. Fig. 4.6 shows the overview of parts of the aggregation protocol (Fig. 4.6).

4.3.3.1 Bootstrapping

In the bootstrapping phase all the sensor nodes are preloaded with master key Key_{IN} which is stored in the nonvolatile storage of the node. When the node boots for the first time, it copies the key from nonvolatile storage to volatile storage and then erases it from the original location. The key Key_{IN} which is available in volatile storage will be removed once the third part, that is, establishment of the key is complete. In the bootstrapping phase the sensor node computes its own master key using methods previously proposed: energy-efficient distributed deterministic key (EDDK) [63], enveloped public key encryption (EPKE) [34], and lightweight extensible authentication protocol (LEAP) [64]. Sensor node A calculates $MasterK_A$ as in Eq. (4.5) with the help of the initial key Key_{IN}. Fig. 4.7 shows the bootstrapping phase of the aggregation protocol.

$$MasterK_A = OWF(Key_{IN}, A_{ID}) \tag{4.5}$$

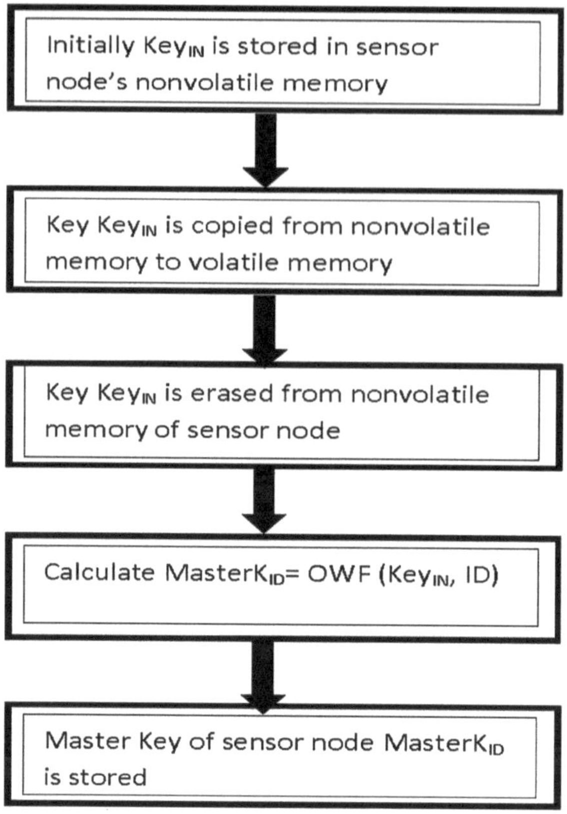

FIGURE 4.7

Bootstrapping phase.

Pseudo code of Bootstrapping Phase:

Algorithm 5: Bootstrapping()

```
Begin
   KeyIN is copied from non-volatile memory to volatile memory.
   Data residing at non-volatile memoryis erased.
   Compute MasterKID= OWF (KeyIN, ID)
   Store MasterKID.
End
```

4.3.3.2 Aggregation tree construction

In the aggregation tree construction phase of the aggregation protocol, the message transferred between the sensor node and its parent contains sequence number, sensor node id, parent node id, and other information in the header. Types of messages used in this part of the protocol are Tree_Beacon, Tree_Join_Request and Tree_Join_Success. In this part the BSN broadcasts the Tree_Beacon message. Fig. 4.8 shows the message exchanges for aggregation tree construction. Fig. 4.8A shows how the base station discovers its child.

$$\text{BSN} \rightarrow {}^{*}\text{:Tree\_Beacon} \tag{4.6}$$

Sensor Node A listens to the beacon message and sends a request via the Tree_Join_Request message to connect with the aggregation tree. Fig. 4.8B shows node A and node B joining the tree.

$$\text{A} \rightarrow \text{BSN:Tree\_Join\_Request} \tag{4.7}$$

Once the BSN receives the Tree_Join_Request message from the sensor nodes, the BSN sends a Tree_Join_Success message as an acknowledgment. The BSN also adds accepted sensor node to its list. BSN can deny further acceptance of child nodes if its defined degree exceeds the degree of tree.

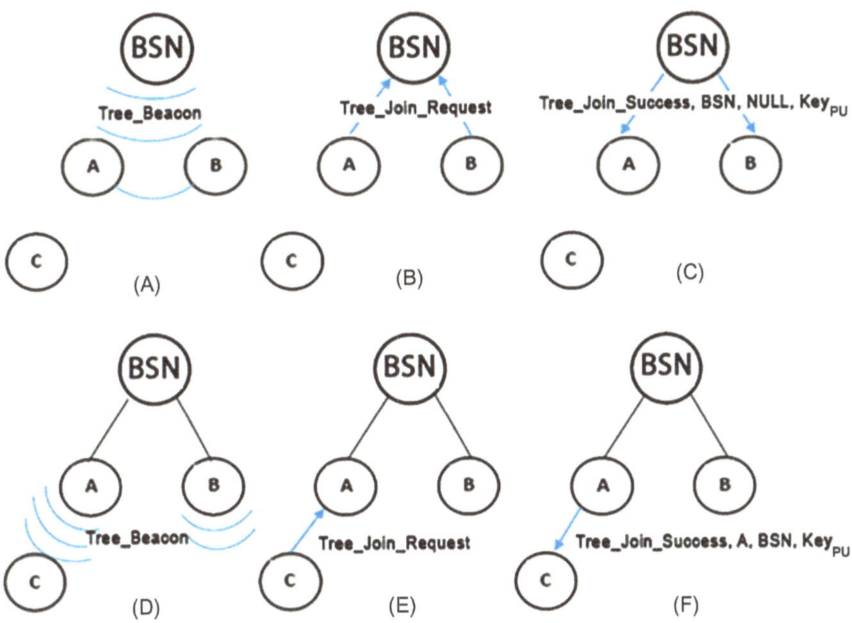

FIGURE 4.8

Tree construction (A) BSN broadcast Tree_Beacon. (B) Sender node A and B send requests to join the tree to the BSN. (C) BSN receives and approves the request by sending the Tree_Join_Success message to sensor node A and B. (D) After discovering their parents, sensor node A and B broadcast Tree_Beacon to allow other nodes to join the tree. (E) Sensor node C attempts to connect to the tree by sending the Tree_join_Request message to node A. (F) Sensor node A receives and approves the request by sending the Tree_Join_Success message.

The Tree_Join_Success message contains two IDs: parent ID and grandparent ID. The sensor node that sends the Tree_Join_Success message changes its own ID to parent ID and changes its grandparent ID to its own parent ID. If the sensor node does not have any grandparent, it sets its grandparent ID to NULL. Fig. 4.8C shows receipt of Tree_Join_Request from any node.

$$\mathrm{BSN} \rightarrow \mathrm{A{:}Tree\_Join\_Success, BSN, NULL, Key_{PU}} \tag{4.8}$$

In Eq. (4.8), the first field is the success message, Tree_Join_Success. The second field is BSN that is the parent for A and the third field is NULL because sensor node A does not have any grandparent. Sensor node A stores its parent information and grandparent information for further communications. The last field is the public key of the BSN that is sent to A. $\mathrm{Key_{PU}}$ is the key used for preserving confidentiality and privacy. After successful joining the tree, sensor node A will broadcast the Tree_Beacon message for other nodes to join the tree as a child of A. Fig. 4.8D shows the way by which sensor node A tries to discover its own children. The equations for node C joining the tree are as follows:

$$\mathrm{A} \rightarrow {*}\mathrm{{:}Tree\_Beacon} \tag{4.9}$$

$$\mathrm{C} \rightarrow \mathrm{A{:}Tree\_Join\_Request} \tag{4.10}$$

$$\mathrm{A} \rightarrow \mathrm{C{:}Tree\_Join\_Success, A_{ID}, BSN, Key_{PU}} \tag{4.11}$$

Once sensor node C has joined the tree and identified its parent and grandparent, node C can send the same messages to other nodes to join the tree as shown in Fig. 4.8E and F. Then the sensor nodes in the aggregation tree will find its parent and grandparent IDs in addition to the public key of the BSN.

Pseudo code of construction of Aggregation Tree is:

Algorithm 6: Tree_Construction_Aggregation

```
Procedure Inititalization()
Begin
    If(BSN)
          broadcasting of<Tree_Beacon> after predefined duration
    End If
End
Procedure recv_Tree_Beacon()
Begin
    Send<Tree_Join_Request>
End
Procedure recv_Tree_Join_Request()
Begin
    If child count <DEGREE_TREE
          Response with message <Tree_Join_Success, ID, Parent, KeyPU>
          Increase value ofchild count by one.
    End If
End
Procedure recv_Tree_Join_Success()
Begin
    Decode received message and take Parent node id and Grandparent node id
    from it.
    Store parent node id and grand parent node id.
    broadcast <Tree_Beacon> after predefined duration
End
```

4.3.4 Establishment of keys

In this part we discuss the key establishment mechanism. The technique used in this section is based on the transitory master key setup protocols EDDK [63], EPKE [34], and LEAP. The Key_Exchange message is transferred from sensor node A to its parent and grandparent. Parent node receives the message and then forwards the message to its own parent and so on. The message which is transferred by A to its parent is as follows:

Sensor node A generates two nonce which are added to the message and then the message is encrypted with the key Key_{IN}. Once the encryption is complete, the MAC is generated on the encrypted message with the same key.

$$\begin{aligned}
&A \rightarrow A_P{:}Key_Exchange, Enc\left(Key_{IN}, A_{N1}||A_{N2}\right)\\
&Mac\left(Key_{IN}, H||Key_Exchange||Enc\left(Key_{IN}, A_{N1}||A_{N2}\right)\right)\\
&A \rightarrow A_{GP}{:}Key_Exchange, Enc\left(Key_{IN}, A_{N1}||A_{N2}\right),\\
&Mac\left(Key_{IN}, H||Key_Exchange||E\left(Key_{IN}, A_{N1}||A_{N2}\right)\right)
\end{aligned} \tag{4.12}$$

When a parent of node A, receives this message, parent node B computes its shared key with node A using Eq. (4.14).

$$MasterK_A = OWF(Key_{IN}, A_{ID}) \tag{4.13}$$

$$KMAC_{A,B} = OWF(MasterK_A \oplus MasterK_B, A_{N1} \oplus B_{N2}) \tag{4.14}$$

Sensor node B replies to the child node once the shared key is calculated with nonce. Hence, the child node can compute the shared key with parent node A. Eq. (4.16) shows the calculation for shared key by the child node. With this calculation, the child node shares the key with its parent and another key with its grandparent.

$$MasterK_B = OWF(Key_{IN}, B_{ID}) \tag{4.15}$$

$$KMAC_{A,GP} = OWF(MasterK_A \oplus MasterK_A, B_{N2} \oplus A_{N1}) \tag{4.16}$$

As XOR shows the commutative property, Eqs. (4.14) and (4.16) compute the same value. Each sensor nodes (child node, parent node, grandparent node) store these calculated keys in a table. At the end of this part of the aggregation protocol, the key table of sensor nodes contains shared keys for all parent, grandparent, child, and grandchild nodes. At the end of the key establishment phase, Key_{IN} is erased from the volatile storage of the sensor node and thus an invader would be unable to access the key or recover Key_{IN} after this phase.

The communications between node A and node P are illustrated in Fig. 4.9. Fig. 4.9A shows how node P and node GP calculate the shared key with node A. The communication between node P, grandparent node GP, and node A is illustrated in Fig. 4.9B. Once nonce is received by node A from node P and node GP, it computes the shared key.

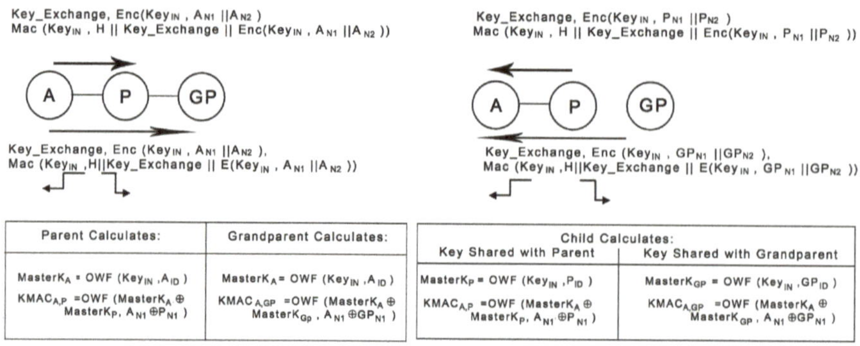

FIGURE 4.9

Key Establishment (A) Parent node and grandparent node receive nonce from child node A. (B) After receiving nonce from parent and grandparent nodes, node A computes shared key with parent and grandparent.

Pseudo code for Establishment of key is:

Algorithm 7: Establishment_Keys

```
Procedure Inititalization()
Begin
    <Key_Exchange, N1_ID, N2_ID> sent to P.
    <Key_Exchange, N1_ID, N2_ID> sent to GP.
End
Procedure recv_Key_Exchange()
Begin
    From message header, source node ID are extracted and stored in temp
    variable R_ID.
    Nonce are fetched and stored in R_N1 and R_N2.
    compute MasterK_R_ID=OWF(Key_IN,R_ID)
    compute KMAC_ID,R_ID=OWF(MasterK_R_ID⊕MasterK_ID, R_N2 ⊕ N2_ID)
    generate new KeySuit={KMAC_ID,R_ID}
    insert new raw as <R_ID, KeySuit> in KeyTable
    Transfer <KeyExchange, N1_ID, N2_ID>to R_ID
    If the Establishment of keys phase is complete
        Delete Key_IN from volatile memory.
    End If
End
```

4.3.5 Aggregation phase

The aggregation function (AF) in this aggregation protocol has the following properties:

AF(V1, V2) = AF(V2, V1): aggregate function is commutative and AF(V1, V2, V3) = AF(V1, AF (V2, V3)): aggregate will be computed using subaggregate values.

The sensor node in the aggregation tree sends its own reading to its parent and grandparent nodes. Each node in the tree sends the same readings to its parent and grandparent nodes, until eventually parent and grandparent node calculate the same aggregate value. Each message in the phase consists of two parts: (1) the sensor node's own reading and (2) subaggregate values that are encrypted with the base station public key Key_{PU}.

Subaggregate values are the aggregate of the values sensed by all the children of a particular node and are based on the aggregation function used. If we use the aggregation functions average or sum, the aggregate value is zero. If we use the aggregation function max, then the subaggregate value is the minimum value from the sensed value. Packet format for packet directed towards parent and grandparent nodes is shown in Fig. 4.10.

$$\begin{aligned} &A \rightarrow A_P{:}H, \text{Parent}, \\ &\text{Enc}\left(Key_{PU}, \text{reading}||\text{sub aggregate}\right), \\ &\text{Mac}\left(\text{MasterK}_{A,P(A)}, H||\text{Parent}||\text{Enc}\left(Key_{PU}, \text{reading}||\text{sub aggregate}\right)\right) \end{aligned} \tag{4.17}$$

The following message is transferred from sensor node A to its grandparent.

$$\begin{aligned} &A \rightarrow A_{GP}{:}\ H, \text{GParent}, \\ &\text{Enc}\left(Key_{PU}, \text{reading}||\text{subaggregate}\right), \\ &\text{Mac}\left(\text{KMAC}_{A,GP(A)}, H||\text{GParent}||\text{Enc}\left(Key_{PU}, \text{reading}||\text{subaggregate}\right)\right) \\ &\text{Mac}\left(\text{KMAC}_{A,P(A)}, H||\text{GParent}||\text{Enc}\left(Key_{PU}, \text{reading}||\text{subaggregate}\right)\right) \end{aligned} \tag{4.18}$$

As mentioned before, every node A shares two keys, K_P and K_{GP}, with its parent and its grandparent. These two keys are used in generating two different MACs, such that verification of the MAC is completed twice by the parent and the grandparent. Such verification is also justified from the point of view that the packet destined to a grandparent is routed through the parent node, which may not be trusted by the grandparent. Thus, the grandparent node exercises its own integrity check on the incoming packet. A node must also aggregate the results before onward transmission of the packet, thus after receiving all the packets PC_1, PC_2, PC_3, . . ., PC_n from all of its child nodes 1 to n, the node verifies the integrity of the incoming MAC from all the packets, aggregates the data values in the packets and attaches a new computed MAC to the outgoing packet.

A node also aggregates the values in the packets it receives from its grandchildren. Thus for any node A, denoting the values received from its child nodes as

header	Reading$_A$	Sub Aggregate$_A$	MAC(With Parent Key)

(A) Packet designated to Parent

header	Reading$_A$	Sub Aggregate$_A$	MAC(With Grand Parent Key)	MAC(With Parent Key)

(B) Packet designated to Grandparent

FIGURE 4.10

Message presentation (Packet) (A) communication to P and (B) communication to GP.

($reading_{G1}$, $reading_{G2}$, ……… $reading_{Gn}$) and denoting the aggregated results received from its grandchildren nodes as ($subaggregate_{GP1}$, $subaggregate_{GP2}$, …………… $subaggregate_{GPm}$), the new aggregated values at the sensor node would be given by subaggregate as shown in Eq. (4.19).

$$subaggregate_A^1 = AF(AF(reading_{G1}, reading_{G2}.\ldots.reading_{Gn}), AF(subaggregate_{G1}, subaggregate_{G2}, \ldots subaggregate_{GPn},)) \quad (4.19)$$

In order to verify the integrity of the aggregated value, the node A would also compute aggregation of all the subaggregated values received from its child nodes. This is shown in Eq. (4.20).

$$subaggregate_A^2 = AF(subaggregate_{P1}, subaggregate_{P2}, subaggregate_{P3}, \ldots\ldots\ldots subaggregate_{Pn}) \quad (4.20)$$

If the two computed values given by Eqs. (4.19) and (4.20) are the same, then the integrity of the message is verified. In all other cases, the node would reject the aggregated values and compute new aggregated values itself, based on the raw values received from its child nodes.

If the value in Eqs. (4.19) and (4.20) are same, then the sensor node computes aggregate values for the sub tree at the current node using Eq. (4.21).

$$subaggregate = AF(subaggregate_{P1}, subaggregate_{P2}, subaggregate_{P3}, \ldots, reading_{P1}, reading_{P2}, reading_{P3}, \ldots) \quad (4.21)$$

After calculating subaggregate values in Eq. (4.21), the sensor node will send all the values to its parent. Then the sensor node will generate its own message for its own parent and grandparent. In Fig. 4.11, the aggregation sum is calculated.

Procedure for aggregation is:

- r_A, r_B, r_D, r_G are the values sensed by node A, B, D, and G, respectively, and encrypted with BS public key Key_{PU}. Node A, B, D, and G set the subaggregate value to zero due to the aggregation function used. Then, as in Fig. 4.11A, each leaf node sends these values to their parent and grandparent.
- Node C has two children, node A and node B. Node C computes $r_A + 0 + r_B + 0$. Node E has only one child, node D and thus computes $r_D + 0$. As in Fig. 4.11B, node C and node E will send this aggregate value and its sensed value to parent and grandparent nodes.
- Node F now receives GParent type of message from the node A, B, and D, and Parent type of message from node C and E. As shown in Fig. 4.11C, node F will do the following computations:
  - Node F calculates $r_A + 0 + r_B + 0 + r_D + 0$ as aggregation of the messages received from nodes A, B, D as a grandchild.
  - Node F then calculates $r_A + 0 + r_B + 0 + r_D + 0$ as aggregation of the messages received from the nodes C and E as a child.

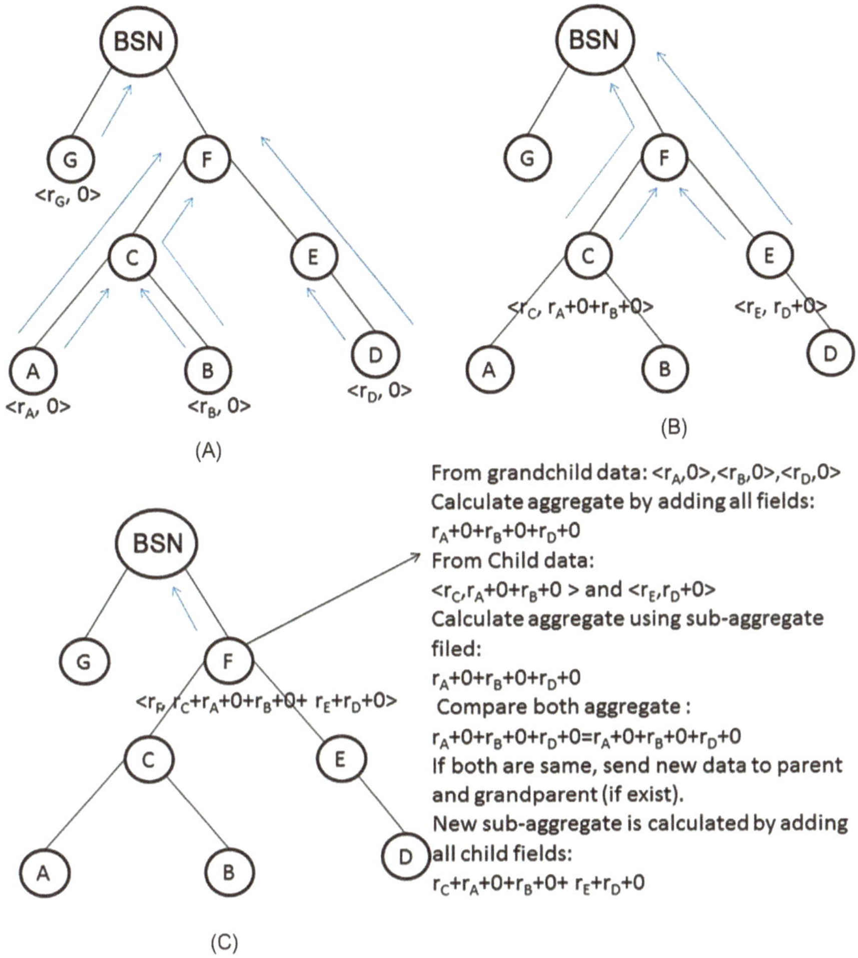

FIGURE 4.11

Aggregation function (AF): (A) Message is transferred to parent and grandparent nodes by leaf node. (B) Parent node receives message from child and applies aggregation function and generates new message for its parent and grandparent. (C) Grandparent also receives message from its child and grandchild. Grandparent again applies aggregation and compares results for integrity checking.

- Node F now compares both the computed values. If both values are the same, it performs further processing. Otherwise it will inform the base station of aggregation value and recover the original aggregate value from GParent type of message.
- Then node F will compute its new aggregation value from all the child nodes C and E. Hence, the new subaggregation value that F computes is: $r_C + r_A + 0 + r_B + 0 + r_E + r_D + 0$.

- Node F then sends $<r_F, r_C + r_A + 0 + r_B + 0 + r_E + r_D + 0>$ to the base station.
- The BSN then performs the same computation as of node F and obtains the final aggregate value from the message of Parent type.
  - Node F and node G send Parent type messages to the BSN and node C and E send GParent type messages to the base station.
  - The BSN performs the decryption with the help of private key Key_{PR} of the BSN and performs the verification.
- The BSN will have the aggregation value from the messages of parent type. Final aggregate value is calculated using the values received from node G and F. The final aggregate value computed is: $r_G + 0 + r_F + r_C + r_A + 0 + r_B + 0 + r_E + r_D + 0$. After receiving this, the BSN will decrypt the aggregate value with its own private key Key_{PR} to preserve privacy of the data.

  Pseudo code of aggregation phase is:

Algorithm 8: Aggregation_Phase

```
Procedure Initialization()
Begin
    If leaf_node
        call send_Parent_Grandparent(INITIAL_SUBAGGREGATE)
    End If
End
Procedure send_Parent_Grandparent(Subaggregate)
Begin
    Parent and GrandparentKeySuite extracted.
    Send message <Parent, Enc(Key_PU, Subaggregate || sense_data),
    Mac (KMAC_ID,Parent, Parent||Subaggregate || sense_data)>
    Send <GParent, Enc(Key_PU, Subaggregate || sense_data)
    Mac (KMAC_A,Grandparent, GParent||Subaggregate || sense_data),
    Mac (KMAC_ID,Parent, GParent||Subaggregate || sense_data)>
End
Procedure aggregation_phase()
Begin
    Message received from children.
    Message received from grandchildren.
    Verification of integrity and authencticy is done by extracting key for
    nodes.
    For non-responding node, alert sent to BSN and node P.
    Reading and sub aggregate values are extracted from all the grand child
    type of nodes and subaggregate_1 is computed as subaggregate_1 = AF (AF
    (reading_G1, reading_G2 ...), AF (subaggregate_G1, subaggregate_G2...))
    Sub aggregate value is extracted from the all the child type of nodes
    and subaggregate_2 is computed assubaggregate_2 = AF(subaggregate_P1,
    subaggregate_P2, subaggregate_P3, ...)
    If subaggregate_1=subaggregate_2
        calculate sub aggregate= AF(subaggregate_1, reading_P1, reading_P2,
        reading_P3, ...)
        call send_Parent_Grandparent(subaggregate)
    Else
        alert sent to the BSN.
        Aggregate valued are calculated using results from immediate
        child nodes
        call send_Parent_Grandparent(subaggregate_1)
    End If
End
```

4.4 Solutions

4.4.1 Example of end-to-end approach

In AdaptableSDA node 3 generates cipher-text of the message of node 3 with the help of the EC-OU encryption function and public key of the BS and hence $C_3(x; y)$ is generated. Node 3 also generates signature S_3 with the help of the proposed variant of EC-DSA, using its own private key and sends its public key Pk_3 on the outgoing message M_3. Node 3 transfers cipher-text $C_3(x; y)$, signature S_3 and public key Pk_3 to its parent node 1. Similarly, Node 4 transfers cipher-text $C_4(x; y)$, signature S_4 and public key Pk_4 to its parent node 1. Once messages are received from node 3 and node 4, node 1 applies aggregation function SUM (elliptic curve point addition) on cipher-texts $C_3(x; y)$ and $C_4(x; y)$ and generates its own cipher-text $C_1(x; y)$. Node 1 also applies addition on received signatures S_3 and S_4 and generates its own signature S_1. In addition, node 1 also applies point addition (elliptic curve point addition) on public keys Pk_3 and Pk_4 and generates its public key Pk_1. Finally, node 1 transfers cipher-text $C_1(x; y)$, signature S_1 and public key Pk_1 to its parent node 0 (BS). Node 2 performs similar steps on the messages received by node 5 and node 6. Node 0 decrypts the message with the help of the EC-OU decryption function and its own private key. Then node 0 would verify the signature with the received public key Pk_0. If the signature is verified, node 0 accepts the message. If the signature is not verified, node 0 discards the message. Thus AdaptableSDA preserves confidentiality, privacy, and data integrity in the tree architecture. Fig. 4.12 shows an example of the AdaptableSDA. The notations used are as follows:

S_i = Signature generated by ith node
C_i = Cipher-text generated by ith node
Pk_i = Public key of ith node

4.4.2 Example of hop-by-hop approach

In AdaptableSDA, node 3 generates cipher-text of the message of node 3 with the help of EC-OU encryption function and public key of the BS and hence, $C_3(x; y)$ is generated. Node 3 also generates MAC_3 on cipher-text with the key shared between the leaf node and the aggregator node. Node 3 transfers cipher-text $C_3(x; y)$ and $MAC_3(x; y)$ to its parent node 1. Similarly, Node 4 transfers cipher-text $C_4(x; y)$ and $MAC_4(x; y)$ to its parent node 1. After receiving messages from node 3 and node 4, node 1 verifies the MAC received by node 3 and 4. If it is verified, node 1 accepts the message and applies aggregation function SUM (elliptic curve point addition) on cipher-texts $C_3(x; y)$ and $C_4(x; y)$ and generates its own cipher-text $C_1(x; y)$ and computes $MAC_1(x; y)$ on it. Then, node 1 transfers cipher-text $C_1(x; y)$ and $MAC_1(x; y)$ to its parent node 0 (BS). Node 2 performs similar steps on the messages received by node 5 and node 6. Node 0 (BS)

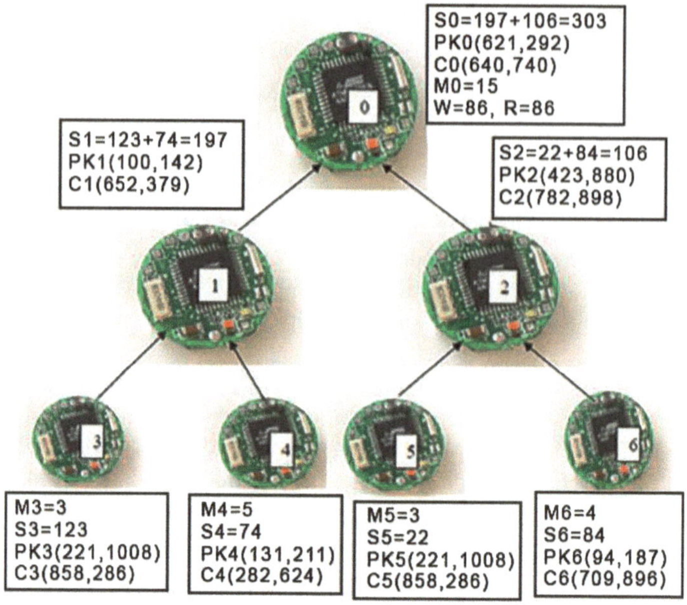

FIGURE 4.12

Example of ECDSA (Bloom filter) based integrity.

verifies the MAC. If the MAC is verified, node 0 decrypts the message with the help of EC-OU decryption function and its own private key and accepts the message. If the MAC is not verified, node 0 simply discards the message. Thus, AdaptableSDA preserves confidentiality, privacy and hop-by-hop data integrity in tree architecture. Fig. 4.13 shows an example [65] of the AdaptableSDA. We used the following representations:

Msg = Sensor node message
MAC_i = MAC calculated with SHA-1 by ith node
C_i = Cipher-text generated by ith node

4.4.3 Metrics of evaluation

As it is known that the sensor nodes in a body area network are themselves resource-constrained, the algorithm employed must be carefully designed to save energy and increase the lifespan of body area network. Typical Mica2 motes have only 4 KB of RAM and 128 KB of flash memory. Hence, to evaluate the AdaptableSDA, we use different metrics viz. energy consumption in Joule and memory requirements (RAM and ROM).

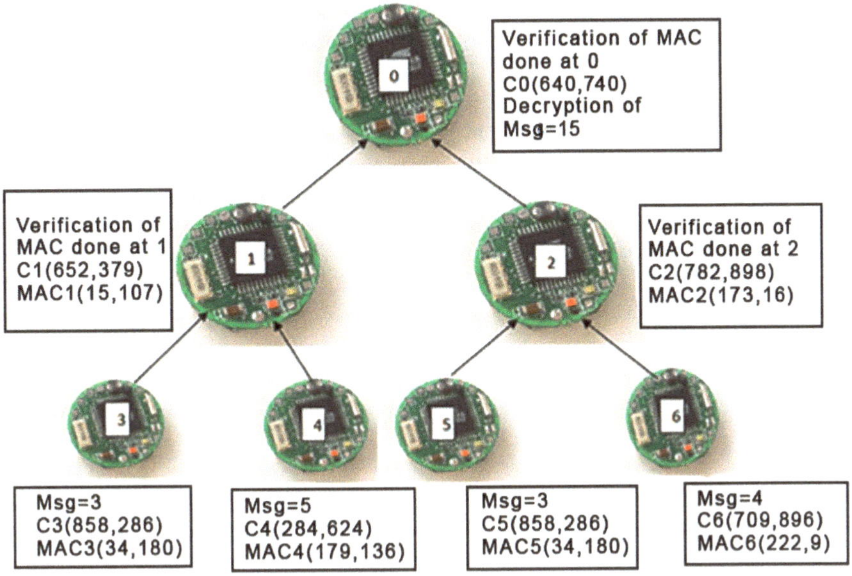

FIGURE 4.13

Example of MAC-based approach.

4.4.4 Results of various configurations of SDA

Table 4.2 shows the experimental results of the approach with various configurations. The first configuration (i.e., confidentiality and privacy preserving SDA requires 7.59% more RAM and 4.45% more ROM compared to the approach without SDA). The approach of SDA supporting ECDSA-based integrity requires 35.02% more RAM and 33.12% more ROM, but provides the required security attributes such as confidentiality, privacy, and integrity. The Bloom filter-based ECDSA approach requires 40.42% more RAM and 36.01% more ROM and provides the required security attributes with increased and intrinsic security. Moreover, the hop-by-hop approach using MAC needs 36.45% more RAM and 20.06% more ROM compared to without SDA.

In the AdaptableSDA configuration of only confidentiality and privacy without integrity requires less resource consumption. Use of a configuration of confidentiality, privacy and EC-DSA based integrity, consumes 34% more resources compared to that of without SDA, due to the inclusion of additional security attributes. In the Bloom filter-based configuration, resource consumption is highest as compared to other configurations of the AdaptableSDA. The inclusion of Bloom-filter data structure requires additional resources, but increases the security strength and is feasible in a body area network. Use of MAC for integrity requires less resource consumption compared to EC-DSA based integrity because MAC

Table 4.2 Percentage of increase compared to without SDA.

Various configurations of AdaptableSDA	RAM in bytes (%)	ROM in bytes (%)	Energy consumption (%)
Confidentiality and privacy reserving SDA	7.59	4.45	0.002
Confidentiality, privacy and ECDSA-based integrity preserving SDA	35.02	33.12	0.008
Confidentiality, privacy and ECDSA (Bloom filter) based integrity preserving SDA	40.42	36.01	0.00
Confidentiality, privacy, and MAC based authentication	36.45	20.06	0.008

Table 4.3 Aggregation protocol memory requirements.

Approach	Platform	RAM (bytes)	ROM (bytes)
Without privacy	MicaZ	1335	36468
	TelosB	1408	28770
With privacy	MicaZ	2081	41572
	TelosB	1996	33362

does not involve expensive computations like signatures. Thus depending on the requirements of an application the discussed configurations of the AdaptableSDA may be utilized in a body area network.

4.4.5 Results of aggregation protocol

In Table 4.3, the memory requirements with and without privacy for the aggregation protocol are outlined. One can observe from the table that the aggregation protocol has 30% overhead in storage, but also provides the important security attribute of privacy. The aggregation protocol has only 9% overhead in energy consumption while providing important security attributes. The comparative energy consumption using MicaZ and TelosB mote when ensuring privacy with our framework is shown in Table 4.3. The figures in the table show the storage requirement in the code as well as data storage for both these motes with and without privacy. It is emphasized that because TelosB uses 16-bit MSP-430 microcontroller as compared to 8-bit AVR microcontroller in MicaZ, there is a perceived increase in storage compared to the MicaZ.

The key establishment in the proposed framework requires special mention with respect to the one proposed in [34]. It is obvious that in order to ensure security, the key used to enable the participating nodes must be deleted once the key setup is completed (at time T_{min}). However, it is essential that within T_{min}, all

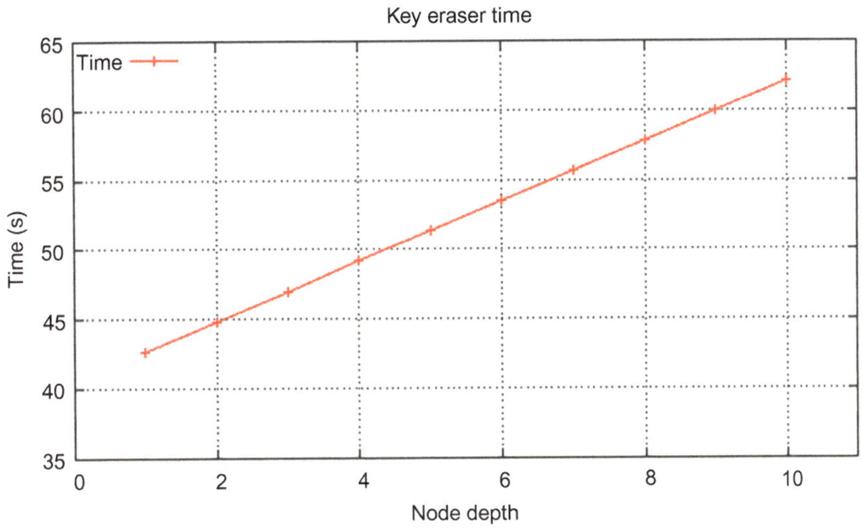

FIGURE 4.14

Key eraser timing at different depth of the tree.

nodes are able to share the key using the initial key. If a node shares the key with all the other nodes in the network, it increases the key storage cost and complexity to $O(n^2)$, while reducing the probability of completing the key setup within the stipulated T_{min} time. To ensure that the key setup is feasible in the proposed approach, nodes share keys with only their parent and its grandparent nodes. In the approach proposed in [34], nodes share keys with all other nodes in the network—reducing the probability of completing the task of key setup within T_{min}. We analyzed the impact of the topology changes on the time T_{min}. The results of varying the depth of the tree topology and its impact on T_{min} are shown in Fig. 4.14.

The experimental results obtained in the current work can be interpreted as follows: if a node in a network is at a depth of 2, then the time required to delete the keys from the nodes at level 2 is 44.88 seconds. This time increases to 62.11 seconds for all those nodes that are at a depth of 10. Thus it can be inferred that the time required to delete the key rises at the rate of 2.15 seconds. Projecting the same for the nodes at the depth 50, the time turns out to be 148.11 seconds, which is less than 2.5 minutes. From this discussion, it can be inferred that an attacker with an objective to recover the key can succeed only in the initial time. However, assuming that an attacker captures a node for the purpose of recovering the key, the probability of success is low recovery of the key from a node requires reading from volatile memory after capturing the node and once the time required for the purpose elapses, the key cannot be recovered.

The experimental evaluation in the current work was carried out on Cooja simulator with a MSPSim plug in [66]. The motes used in simulation were the TelosB nodes. In order to obtain the estimate of the energy consumed by the TelosB motes, the approach discussed by the authors in [67] and [66] was employed. The simulation results obtained in this manner, indicate that, an average of 35% more energy is consumed during the aggregation phase as compared to the setup phase (Fig. 4.15). The reason for this can be attributed to the fact that the aggregation phase involves additional operations that yield the security and privacy attributes.

The experimental results for total energy consumption (i.e., the energy consumed during the setup phase as well as during the aggregation phase) for TelosB motes with support for privacy is shown in Fig. 4.16. As indicated in the figure, the leaf nodes entail less energy expenditure because the aggregation operation is performed only by the intermediate nodes.

An estimate of the penalty due to the cryptographic operations is illustrated in Fig. 4.17. It is clear that the protocol preserves privacy with insignificant additional overhead of less than 10%.

Table 4.4 displays the increased number of security operations to be performed by the grandparent node. There is an increase in energy consumption of less than 32%.

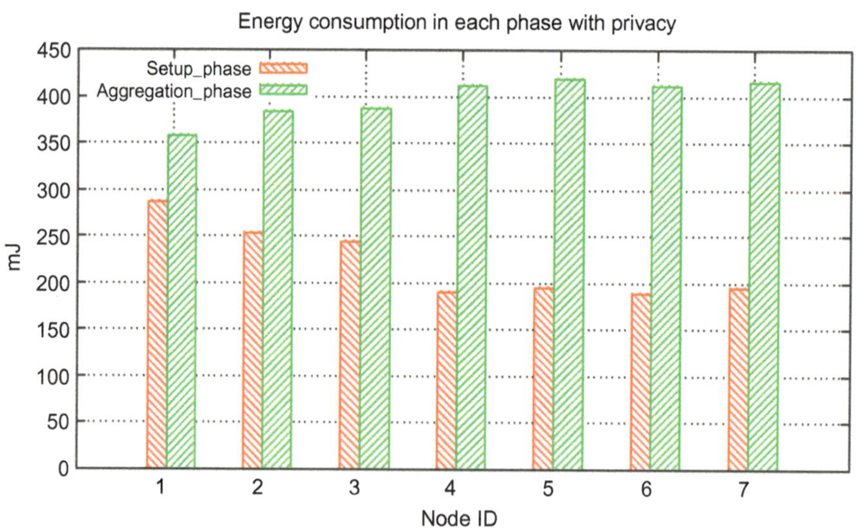

FIGURE 4.15

Energy consumption in setup phase and aggregation phase with privacy preservation of Aggregation Protocol for TelosB motes.

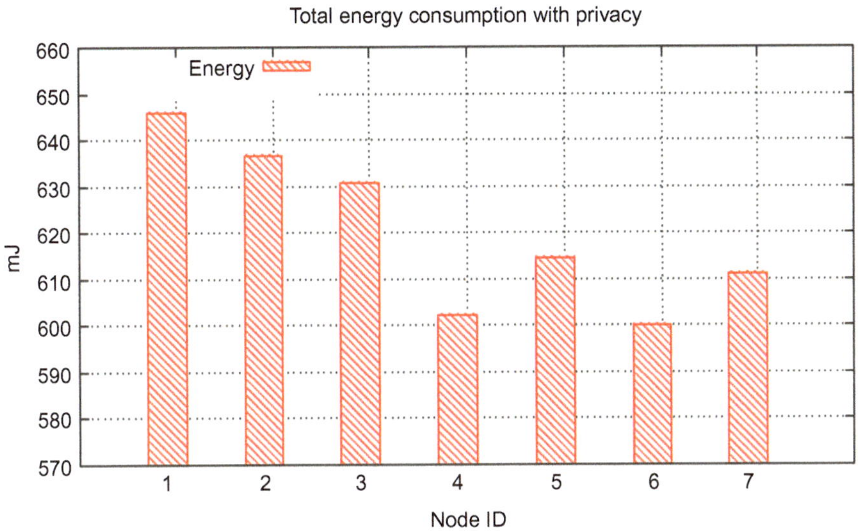

FIGURE 4.16

Total energy consumption with privacy preservation of the aggregation protocol for TelosB motes.

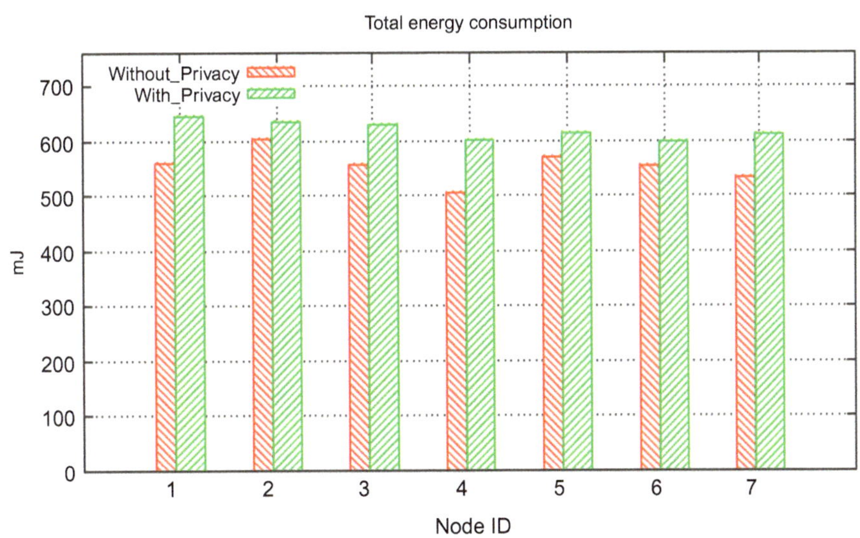

FIGURE 4.17

Energy consumption in the aggregation protocol with and without cryptographic primitives.

Table 4.4 Comparing the energy consumption of a parent and a grandparent node.

Type	Energy (μJoule)
Packet for parent (generation and sending)	1981.26
Packet for grandparent (generation and sending)	2883.01

4.5 Future research directions

4.5.1 Existing issues, challenges, and problems

Inherent features of a body area network consisting of sensors introduce unique challenges for SDA. In addition, one cannot directly employ existing protocols for data aggregation in a body area network due to the following constraints:

- Uncontrolled environment: sensor nodes are often deployed in remote environments, hence it becomes easy for an adversary to launch physical attacks to capture a sensor node itself and retrieve information stored on it.
- Resource constraints in the sensor nodes: sensor nodes have resource constraints (battery, lifetime) that prohibit application of existing protocols to sensor nodes. For example, fully homomorphic encryption would have higher resource overhead and accordingly may lessen WSNs lifetime.
- Constraints in topology: A body area network may have a large number of nodes deployed that transfer data from sensor nodes to the BS. Body area networks generally exhibit an unbalanced network traffic pattern, which must be accounted for in protocol design.

Considering these issues and challenges in the design of a protocol for SDA, a high level of security in the aggregation process by providing support for the required security attributes is essential. Hence, a prospective malicious insider node is not able to inject false data during the aggregation process and the system offers higher throughput.

4.5.2 Future directions

In future, this work can be extended by exploring alternatives to fully homomorphic encryption algorithms for SDA in body area networks. Moreover, this research can be extended for generation of diverse and dynamic topologies depending on the requirement of real time applications. In addition, this research can be extended for the module of key management where key management methodology would deal with more dynamic topologies. This can also be extended by deploying the complete AdaptableSDA on MicaZ and TelosB motes to make it directly useful for the applications discussed.

The primary factors that have a major impact on protocols of body area networks that use sensors are resource limitations, deployment of sensor devices in hostile environments, and lack of tamper-proof hardware for sensor devices. Due to these factors, there are many open research issues in the field of body area networks. In addition to the future work discussed above, researchers can aim for a comprehensive framework of SDA by considering following points:

- Support of the required security attributes (i.e., privacy, confidentiality, integrity, authentication, freshness, and availability of the data).
- Intrusion detection mechanisms for SDA.
- Key management module for SDA.
- Topology formation for SDA.

4.6 Conclusion

The WBANs have long exhibited tremendous potential in real-world applications such as health care monitoring systems. Examples in healthcare monitoring include assistance to elderly people at home, monitoring of blood sugar and pressure, rehabilitation after surgery, and biofeedback applications.

The WBAN nodes are constrained in resources, that is, computational, storage, battery and bandwidth. Sophisticated techniques have been used to conserve the battery power of the sensor nodes in order to extend the operating lifetime of the sensor nodes. One such technique is on-the-fly processing (aka in-network processing) of data packets in the network itself. In this chapter, we proposed a novel SDA framework—AdaptableSDA for WBANs that uses homomorphic encryption and appropriate data integrity (end-to-end and hop-by-hop) mechanisms to offer confidentiality, privacy, and data integrity attributes. We carried out extensive simulations, experimentations, and detailed analyses in order to justify the AdaptableSDA. They conclude that the AdaptableSDA in WBANs indeed helps in fine-tuning security attributes, thereby minimizing the resource overhead and attaining optimal performance. We also presented an efficient tree topology construction mechanism for SDA in WBANs. They present a lightweight, tree architecture-based key management system for WBANs. The novelty of this work lies in the erasure of the master key in less than 2.5 minutes. Hence, in key distribution, information stored in the device would be inaccessible to an adversary after the initial 2.5 minutes, and thus this approach provides security to the SDA approach.

References

[1] S. Movassaghi, M. Abolhasan, J. Lipman, D. Smith, A. Jamalipour, Wireless body area networks: a survey, IEEE Commun. Surv. Tutor. 16 (3) (2014) 1658–1686.

[2] R. Negra, I. Jemili, A. Belghith, Wireless body area networks: applications and technologies, Proc. Comp. Sci 83 (2016) 1274–1281.

[3] S. Al-Janabi, I. Al-Shourbaji, M. Shojafar, S. Shamshirband, Survey of main challenges (security and privacy) in wireless body area networks for healthcare applications, Egypt. Inform. J 18 (2) (2017) 113–122.
[4] N. Dey, et al., Watermarking in biomedical signal processing, Intelligent Techniques in Signal Processing for Multimedia Security, 660, Stud. Comput. Intell. 2017, pp. 345–369.
[5] S. Chakraborty, S. Chatterjee, N. Dey, A.S. Ashour, A.E. Hassanien, Comparative approach between singular value decomposition and randomized singular value decomposition-based watermarking, in: N. Dey, V. Santhi (Eds.), Intelligent Techniques in Signal Processing for Multimedia Security, 660, Stud. Comput. Intell. 2017, pp. 133–149.
[6] N. Dey, M. Dey, S.K. Mahata, A. Das, S.S. Chaudhuri, Tamper detection of electrocardiographic signal using watermarked bio-hash code in wireless cardiology, Int. J. Sign. Imag. Syst. Eng. (IJSISE) 8 (No. ½) (2015) 46–58.
[7] N. Dey, S. Samanta, X.-S. Yang, A. Das, S.S. Chaudhuri, Optimisation of scaling factors in electrocardiogram signal watermarking using cuckoo search, Int. J. Bio-Insp. Comput. (IJBIC) 5 (5) (2013) 315–326.
[8] J. Luprano, J. Sola, S. Dasen, J.M. Koller, O. Chelelat, Combination of body sensor networks and on-body signal processing algorithms: the practical case of MyHeart project, in: Proceedings of International Workshop on Wearable and Implantable Body Sensor Networks (BSN 2006), Cambridge, MA, 2006.
[9] P. Khan, Md. A. Hussain, K.S. Kwak, Medical applications of wireless body area networks, Int. J. Digit. Cont. Technol. Appl. 3 (3) (2009) 185–193.
[10] R. Rajagopalan, P.K. Varshney, Data aggregation techniques in sensor networks: a survey, Comm. Surv. Tutor. IEEE 8 (2006) 48–63.
[11] J. López, J. Zhou (Eds.), Wireless Sensor Network Security, vol. 1, IOS Press, Netherlands, 2008.
[12] F. Akyildiz, W. Su, Y. Sankarasubramaniam, E. Cayirci, Wireless sensor networks: a survey, Comp. Network. 38 (4) (2002) 393–422.
[13] F. Josep, Domingo, A new privacy homomorphism and applications, Inf. Process. Lett. 60 (5) (1996) 277–282.
[14] T. Wang, X. Qin, L. Liu, An energy-efficient and scalable secure data aggregation for wireless sensor networks, Int. J. Distrib. Sens. Netw. (2013). Volume 2013, Article ID 843485, 1-11, 2013.
[15] A. Jangra, Priyanks, Richa, Cb-SDA: cluster-based secure data aggregation for private data in WSN, Wireless Mobile Technol. Sci. Edu. Publ. 1 (1) (2013) 37–41.
[16] H. Li, K. Lin, K. Li, Energy-efficient and high-accuracy secure data aggregation in wireless sensor networks, J. Comput. Commun. 34 (4) (2011) 591–597.
[17] S. Peter, D. Westhoff, C. Castelluccia, A survey on the encryption of converge cast-traffic with in-network processing, IEEE Trans. Depend. Secure Comput. 7 (1) (2010) 20–34.
[18] J.M. Bahi, C. Guyeux, A. Makhoul, Efficient and robust secure aggregation of encrypted data in sensor networks, Proceedings of the fourth International Conference on Sensor Technologies and Applications (SENSORCOMM '10), IEEE Computer Society, Washington, DC, 2010, pp. 472–477.
[19] A.S. Poornima, B.B. Amberker, SEEDA: Secure end-to-end data aggregation in wireless sensor networks, in: Proceeding of the7th International Conference of Wireless and Optical Communications Networks (WOCN), Colombo, 2010, pp. 1–5.

[20] X. Wang, J. Li, X. Peng, B. Zou, Secure and efficient data aggregation for wireless sensor networks, Proceeding of Vehicular Technology Conference Fall (VTC 2010-Fall), IEEE, Ottawa, ON, 2010, pp. 1–5. 2010 72nd.

[21] S. Ozdemir, Y. Xiao, Hierarchical concealed data aggregation for wireless sensor networks, in: Proceedings of Embedded Systems and Communications Security Workshop in conjunction with IEEE SRDS, 2009, pp. 27–29.

[22] R. Bista, K.-J. Jo, J.-W. Chang, A new approach to secure aggregation of private data in wireless sensor networks, in: Proceedings of the Eighth IEEE International Conference on Dependable Autonomic and Secure Computing, 2009, pp. 394–399.

[23] A. Liu, P. Kampanakis, P. Ning, TinyECC: elliptic curve cryptography for sensor networks, in: Proceedings of International Conference on Information Processing in Sensor Networks, St. Louis, MO, 2008, pp. 245–256.

[24] M. Oenen, R. Molva, Secure data aggregation with multiple encryption, in: Proceedings of Fourth European Conference on Wireless Sensor Networks (EWSN 2007), 2007.

[25] O. Ugus, A. Hessler, D. Westhoff, Performance of Additive Homomorphic EC-ElGamal Encryption for TinyPEDS, in: GI/ITG KuVSFachgespräch, DrahtloseSensornetze, RWTH Aachen, 2007.

[26] X. Huang, M. Yang, Y. Tong, An efficient and secure aggregation of encrypted data for wireless sensor network based on dynamic cluster, Proceedings of the simulation multi conference—Volume 1 (SpringSim '07), Society for Computer Simulation International, San Diego, CA, 2007, pp. 51–57.

[27] W. He, X. Liu, H. Nguyen, K. Nahrstedt, T. Abdelzaher, PDA: Privacy-preserving data aggregation in wireless sensor networks, in: Proceedings of the IEEE Infocom 2007, 2007, pp. 2045–2053.

[28] H. Çama, S. Özdemira, P. Nairb, M. Devasenapathy, H.O. Sanlia, Energy-efficient secure pattern based data aggregation for wireless sensor networks, J. Comp. Commun. Arch 29 (4) (2006).

[29] Girao J., Westhoff D., Schneider M. (2005). CDA: Concealed data aggregation for reverse multicast traffic in wireless sensor networks, in: Proceedings of IEEE International Conference on Communications, ICC 2005, vol. 5, pp. 3044–3049.

[30] M. Rezvani, A. Ignjatovic, E. Bertino, S. Jha, Secure data aggregation technique for wireless sensor networks in the presence of collusion attacks, IEEE Trans. Depend. Secure Comput. 12 (1) (2015) 98–110.

[31] S. Roy, M. Conti, S. Setia, S. Jajodia, Secure data aggregation in wireless sensor networks: filtering out the attacker's impact, IEEE Trans. Inform. Foren. Security 9 (4) (2014) 681–694.

[32] S.M. Rafi, K. Subbarao, Secure data aggregation in wireless networks, Int. J. Res. Comp. Commun. Technol. 3 (1) (2014) 87–93.

[33] L. Shen, J. Ma, X. Liu, F. Wei, M. Miao, A secure and efficient id-based aggregate signature scheme for wireless sensor networks, 2012.

[34] M. Bagaa, Y. Challal, A. Ouadjaout, N. Lasla, N. Badache, Efficient data aggregation with in network integrity control for WSN, J. Paral. Distrib. Comput. 72 (10) (2012) 1157–1170.

[35] J. Albath, S. Madria, Secure hierarchical data aggregation in wireless sensor networks, in: Proceeding of the IEEE conference on Wireless Communication and Networking, 2009, pp. 2420–2425.

[36] H. Alzaid, E. Foo, J.M.G. Nieto, D. Park, A taxonomy of secure data aggregation in wireless sensor networks, Int. J. Commun. Netw. Distrib. Syst. 8 (1) (2008) 101–148.
[37] S. Ozdemir, Functional reputation based reliable data aggregation and transmission for wireless sensor networks, J. Comput. Commun 31 (17) (2008) 3941–3953.
[38] B. Przydatek, D. Song, A. Perrig, SIA: secure information aggregation in sensor networks, J. Comp. Secur 15 (1) (2007) 69–102. Special Issue on Security of Ad-hoc Networks.
[39] K. Wu, D. Dreef, B. Sun, Y. Xiao, Secure data aggregation without persistent cryptographic operations in wireless sensor networks. In: Proceedings of the 25th IEEE International Performance Computing and Communications Conference, IPCCC2006, Phoenix, AZ, 2006.
[40] W. Zhang, S.K. Das, Y. Liu, A trust based framework for secure data aggregation in wireless sensor networks, in: Proceeding of theSensor and Ad Hoc Communications and Networks 2006. SECON '06. 3rd Annual IEEE Communications Society on, vol. 1, 2006, pp. 60–69.
[41] L. Hu, D. Evans, Secure aggregation for wireless networks, Proceedings of the Symposium on Applications and the Internet Workshops (SAINT'03 Workshops), IEEE Computer Society, Washington, DC, 2003.
[42] W. Du, J. Deng, Y.S. Han, P.K. Varshney, A witness-based approach for data fusion assurance in wireless sensor networks, in: Proceedings of the IEEE Global Telecommunications Conference, 2003, pp. 1435–1439.
[43] P.R. Vamsi, K. Kant, Secure data aggregation and intrusion detection in wireless sensor networks, 2015 International Conference on Signal Processing and Communication (ICSC), IEEE, 2015, pp. 127–131.
[44] M. Sardaraz, M. Tahir, A.A. Ikram, SDAF: a secure data aggregation framework for wireless sensor networks, Int. J. Comp. Electr. Eng. 5 (5) (2013) 447.
[45] S. Ozdemir, Y. Xiao, Integrity protecting hierarchical concealed data aggregation for wireless sensor networks, Comput. Network. 55 (8) (2011) 1735–1746.
[46] L. Bhasker, Genetically derived secure cluster-based data aggregation in wireless sensor networks, IET Inform. Secur. 8 (1) (2014) 1–7.
[47] S.B. Othman, A. Trad, H. Youssef, H. Alzaid, Secure data aggregation in wireless sensor networks, in: Ad Hoc Networking Workshop (MED-HOC-NET), 2013 12th Annual Mediterranean, IEEE, 2013, pp. 55–58.
[48] Y. Lu, X. Feng, K. Yi, J. Wan, Secure data aggregation based on interval analysis for wireless sensor networks, Intelligent Human-Machine Systems and Cybernetics (IHMSC), 2014 Sixth International Conference on, vol. 2, IEEE, 2014, pp. 305–308.
[49] L. Wang, L. Chen, J. Huang, Secure data aggregation scheme based on key vector sharing in wireless sensor network, Information and Communications Technologies (ICT 2014), 2014 International Conference on, IET, 2014, pp. 1–6.
[50] D.E. Boubiche, S. Boubiche, A. Bilami, H. Toral-Cruz, Toward adaptive data aggregation protocols in wireless sensor networks, Proceedings of the International Conference on Internet of Things and Cloud Computing, ACM, 2016, p. 73.
[51] V. Kumar, S. Madaria, Performance analysis of secure hierarchical data aggregation in wireless sensor networks, in: Proceedings of theMobile Data Management, 2010, pp. 299–300.
[52] S. Ozdemir, Y. Xiao, Polynomial regression based secure data aggregation for wireless sensor networks, in: Global Telecommunications Conference (GLOBECOM 2011), IEEE, 2011, pp. 1–5.

[53] K.A. Shim, C.M. Park, A secure data aggregation scheme based on appropriate cryptographic primitives in heterogeneous wireless sensor networks, IEEE Trans. Paral. Distrib. Syst 26 (8) (2015) 2128–2139.

[54] H. Hayouni, M. Hamdi, Secure data aggregation with homomorphic primitives in wireless sensor networks: a critical survey and open research issues, 2016 IEEE 13th International Conference on Networking, Sensing, and Control (ICNSC), IEEE, 2016, pp. 1–6.

[55] W. Min, C. Ruixiang, H. Shunbin, A secure data aggregation approach in hierarchical wireless sensor networks, Proceedings of the 10th International Conference on Ubiquitous Information Management and Communication, ACM, 2016, p. 89.

[56] K. Parmar, D.C. Jinwala, Concealed data aggregation in wireless sensor networks: a comprehensive survey, Comput. Netw. 103 (2016) 207–227.

[57] B. Schneier, Applied Cryptography: Protocols, Algorithms, and Source Code in C, John Wiley & Sons, 2007.

[58] R. Roman, J. Zhou, J. Lopez, On the security of wireless sensor networks, Proceedings of Computational Science and Its Applications, ICCSA 2005, Springer, 2005, pp. 681–690.

[59] A. Khalique, K. Singh, S. Sood, Implementation of elliptic curve digital signature algorithm, Int. J. Comp. Appl. 2 (2) (2010) 21–27.

[60] J. Sen, A survey on wireless sensor network security, Int. J. Commun. Netw. Inform. Secur. 1 (2) (2009).

[61] M. Challa, M. Reddy, P. Reddy, J.N. Mungara, et al., Secured public key authentication and energy efficient mac implementation in wireless sensor networks, Int. J. Comp. Sci. Netw. Secur. 12 (8) (2012).

[62] A.S. Wander, N. Gura, H. Eberle, V. Gupta, S.C. Shantz, Energy analysis of public-key cryptography for wireless sensor networks, in: Proceedings of the Third IEEE International Conference on Pervasive Computing and Communications, PerCom2005, 2005, pp. 324–328.

[63] X. Zhang, J. He, Q. Wei, EDDK: energy efficient distributed deterministic key management for wireless sensor networks, EURASIP J. Wireless Commun. Netw (2011).

[64] S. Zhu, S. Setia, S. Jajodia, LEAP: efficient security mechanisms for large-scale distributed sensor networks, Proceedings of the 10th ACM Conference on Computer and Communications Security, ACM, 2003, pp. 62–72.

[65] V.J. Jariwala, D.C. Jinwala, Privacy and integrity preserving secure data aggregation in wireless sensor networks, Proceedings of 11th International Workshop on Security in Information Systems, by SciTePress, Lisbon, Portugal, 2014, pp. 49–60. April, 2014.

[66] J. Eriksson, A. Dunkels, N. Finne, F. Osterlind, T. Voigt, MSPSim–an extensible simulator for msp430-equipped sensor boards, in: Proceedings of the European Conference on Wireless Sensor Networks (EWSN), Poster/Demo Session, 2007.

[67] A. Dunkels, F. Osterlind, N. Tsiftes, Z. He, Software-based on-line energy estimation for sensor nodes, in: Proceedings of the 4th Workshop on Embedded Networked Sensors, 2007.

Further reading

A. Mahimkar, SecureDAV: a secure data aggregation and verification protocol for sensor networks, in: Proceedings of the IEEE Global Telecommunications Conference, 2004, pp. 2175–2179.

A. Bouazizi, G. Zaibi, M. Samet, A. Kachouri, Wireless body area network for e-health applications: overview, in: Proceedings of International Conference on Smart, Monitored and Controlled Cities (SM2C), Sfax, 2017, pp. 64–68.

C. Castelluccia, E. Mykletun, G. Tsudik, Efficient aggregation of encrypted data in wireless sensor networks, Proceedings of the MobiQuitous 2005, IEEE Computer Society, 2005, pp. 109–117.

E. Mykletun, J. Girao, D. Westhoff, Public key based cryptoschemes for data concealment in wireless sensor networks, in: Proceeding of IEEE International Conference on Communications ICC2006, Istanbul, 2006, pp. 2288–2295.

H. Hayouni, M. Hamdi, A novel algorithm for securing data aggregation in wireless sensor networks, International Conference on Ad-Hoc Networks and Wireless, Springer International Publishing, 2016, pp. 263–276.

H. Junru, The improved elliptic curve digital signature algorithm, Proceedings of International Conference on Electronic & Mechanical Engineering and Information Technology, IEEE, Harbin, Heilongjiang, China, 2011, pp. 257–259.

H.-Z. Liao, Y.-Y. Shen, On the elliptic curve digital signature algorithm, Tunghai Sci. 8 (2006) 109–126.

Mica2:Wirelessmeasurement system. Available [Online]: <http://xbow.com/Product_pdf_files/Wireless_pdf/MICA2_Datasheet.pdf>.

P. Paillier, Trapdooring discrete logarithms on elliptic curves over rings, in: T. Okamoto (Ed.), Proceedings of Annual International Conference on Theory and Application of Cryptology and Information Security (ASIACRYPT '00), Springer-Verlag, London, UK, 2000, pp. 573–584.

Y. Yang, X. Wang, S. Zhu, G. Cao, SDAP: a secure hop-by-hop data aggregation protocol for sensor networks, ACM Trans. Inf. Syst. Secur. 11 (4) (2008) 1–43.

CHAPTER

5

Screening and early identification of microcalcifications in breast using texture-based ANFIS classification

K. Sujatha[1], G. Durgadevi[2], K. Senthil Kumar[3], V. Karthikeyan[1], R.S. Ponmagal[4], Rajeswari Hari[5], N.P.G. Bhavani[6], V. Srividhya[6] and Su-Qun Cao[7]

[1]*Department of EEE, Dr. MGR Educational and Research Institute, Chennai, India*

[2]*Department of ECE, New Prince Shri Bhavani College of Engineering and Technology, Chennai, India*

[3]*Department of ECE, Dr. MGR Educational and Research Institute, Chennai, India*

[4]*Department of CSE, Dr. MGR Educational and Research Institute, Chennai, India*

[5]*Department of IBT, Dr. MGR Educational and Research Institute, Chennai, India*

[6]*Department of EEE, Meenakshi College of Engineering, Chennai, India*

[7]*Faculty of Mechanical and Materials Engineering, Huaiyin Institute of Technology, Huaian Shi, China*

Chapter Outline

Wearable and Implantable Medical Devices. DOI: https://doi.org/10.1016/B978-0-12-815369-7.00005-7

5.1 Introduction

Breast cancer is a key cause of death among women, particularly in developed nations, and is the second most common malignancy [1,2]. The World Health Organization's International Agency for Research on Cancer in Lyon, France, reports that that more than 150,000 women die of breast cancer each year (Detounis, 2014). The mortality rates of breast cancer are shown in Fig. 5.1.

The mortality and the frequency rate of breast malignancy were evaluated per 100,000 women around the world [3,4]. In India, out of the total women community, 23% are suffering from Breast cancer, 17.5% by cervical cancer in the major cities, such as Mumbai, Calcutta, and Bangalore. Despite the fact that the frequency is lower in India than in other developed nations, the number of breast cancers cases in India is disturbing [2].

According to the International Agency for Research on Cancer, roughly 89,000 women each year were diagnosed with breast cancer in India in 2004 and 92,000 women in 2012–14 [5,6]. The mortality rates of breast cancer in India are shown in Fig. 5.2.

The mortality and the rate of breast cancer were evaluated per 100,000 women in India in order to identify how many women are suffering from breast cancer.

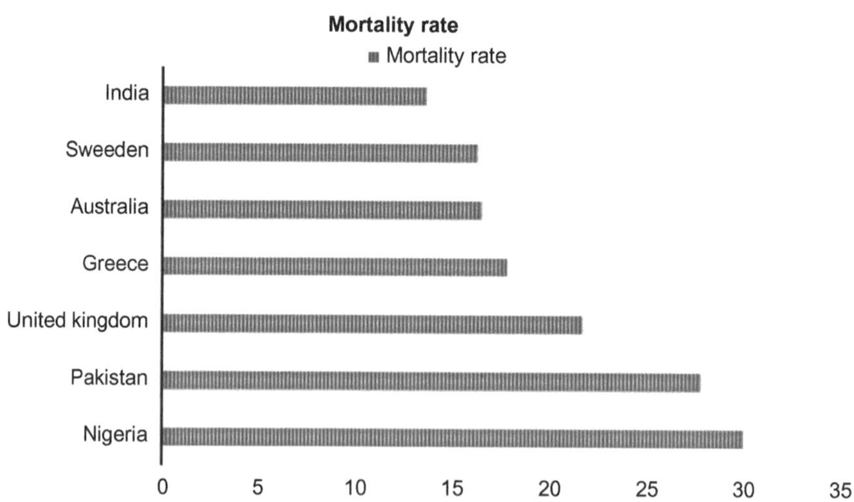

FIGURE 5.1

Mortality rates of breast cancer.

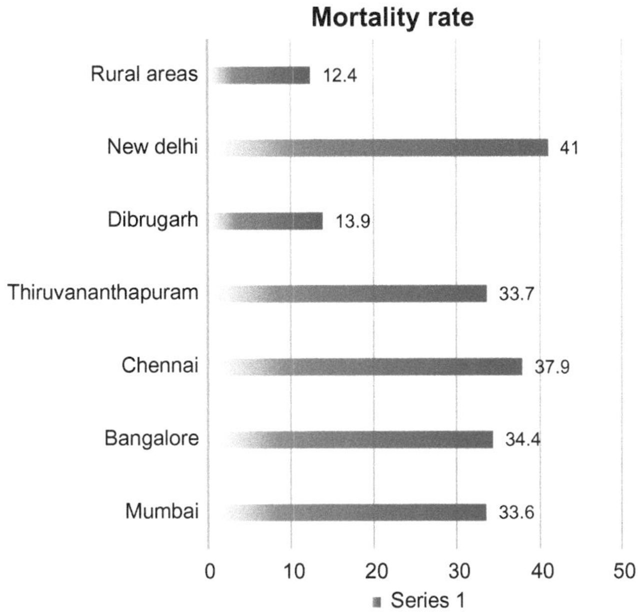

FIGURE 5.2

Mortality rates of breast cancer in India.

The rate is more among urban than rustic women [7,8]. The normal occurrence rate changes from 22 to 28 for every 100,000 women for each year in urban settings to 6 for each 100,000 women for each year in rustic zones. The WHO study reports that by 2020 there will be 10 million new malignancy cases, of which 6 million women will die [9,10]. In India alone 1.5 million new cases of malignancy will be diagnosed during the beginning of this century. Presently screening mammography is supported for all Indian women.

To diagnose breast cancer early, mammography is needed [11,12]. A mammogram is a type of breast X-ray that is used to identify tumor. Figs. 5.3 and 5.4A and B show normal and abnormal mammograms, respectively. A standard mammogram screening is comprised of four images, two perspectives of each breast. Presently screening by mammography is recommended for all Indian ladies.

A radiologist reviews the mammogram images to look for signs of tumors or abnormal growths. A small sized calcification indicates calcium storage in the breast and it is represented by a spot on the image. A group commonly characterized to be no less than 3–5 small-scale calcifications inside a square centimeter distance. Up to 50% of masses show bunched miniaturized scale calcifications and in various cases, the groups are the main indication of danger [13,14].

Calcifications change in size ranging from a smaller one having 0.1 to 5 mm breadth and a radiologist should be able to precisely analyze those calcifications

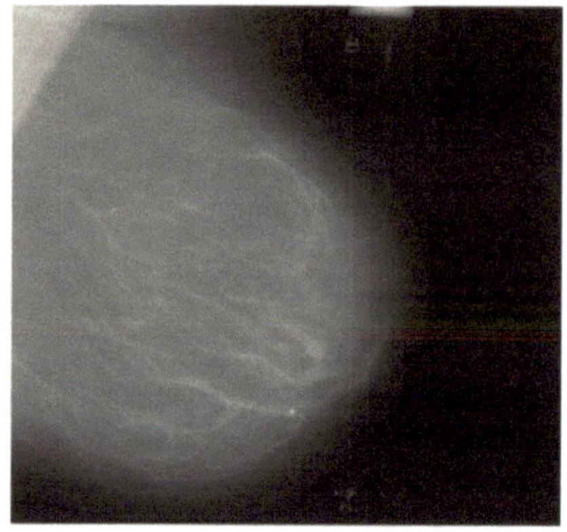

FIGURE 5.3

Normal mammogram.

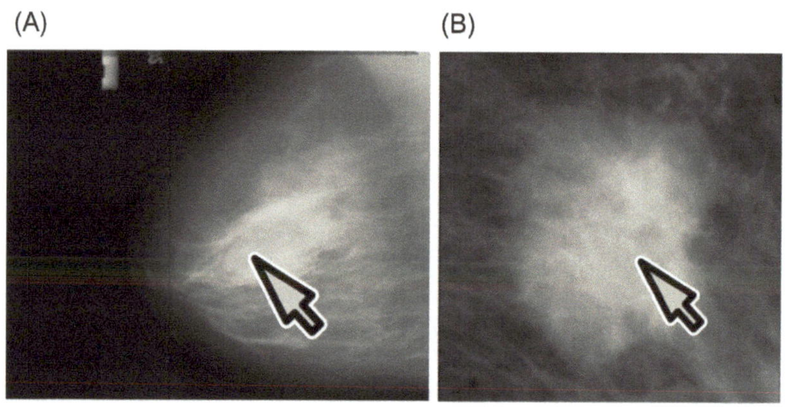

FIGURE 5.4

(A) Presence of malignant tumor in abnormal mammogram. (B) Abnormal mammogram in benign category.

from mammogram by using a magnifier which might be implanted in thick parenchyma (connective) tissue [15,16]. Size, shape, and the thickness of the calcifications vary from person to person and proves to be a complicated problem while diagnosis [17,18].

5.2 Literature review

This section gives an overview of medical imaging modalities and mammographic image classification methods.

5.2.1 Medical imaging modalities

There are various types of medical imaging modalities such as CT, MRI, ultrasound, and mammography used to look for abnormalities such lesions, tumors, microcalcifications, etc., in the human body.

Some researchers have taken support of neural systems in recognizing the patient's encompassed sores from digitized mammograms. It is noted how computerized mammograms are used as a part of single- and multilayer recognition of masses [19].

It is presented as an electronic method of recognizing smaller scale calcification in computerized mammograms. This mode takes a shot at variety in image in which the suppressed image is subtracted from an upgraded image, which evacuates the organized foundation in the mammogram. For removing small-scale calcification signals, edge value–based methods are used. An information-based strategy is reported for recognizing smaller scale calcifications in computerized mammograms.

Bayesian image examination is based on realistic models and general systems. Some people have used filter banks in distinguishing the straight and nodular regions. The subimages were created with the components of a Hessian grid supported by filter banks. The Eigen qualities were ascertained and another filter bank with three properties are used. These properties include (1) nodular regions that are improved with different sizes, (2) different sizes can be found out in both nodular and straight types, and (3) by expelling, a unique image can be reconstructed. In mammograms, a filter bank is used to improve small-scale calcifications [20].

It is suggested that a CAD framework can be used for small-scale calcification identification. In the first step, wavelet and dark level factual properties are used. As a part of small-scale calcification of pixels the small portions are combined together to build them up into objects of individual calcifications. In the second step, 31 factual properties are used to check these potential calcifications. The identification is done using neural networks.

In the work done for early detection of breast cancer using image processing algorithms such as image analysis, binarization, thinning, and Euclidean distance classifiers; the benign and malignant tumors were detected. An accuracy of 99.6% was achieved [22]. To distinguish between benign and malignant tumors, a neural network–based classifier was used [23]. A simple approach using averaging, thresholding, max-mean, and least variance methods are used to detect breast cancer from mammograms are also proposed by various researchers [24].

5.2.2 Mammographic image classification

This section deals with categorizing the acquired images using the imaging modalities discussed in Section 2.1 using techniques like Euclidean classifier, Baysien classifier, back propagation algorithm (BPA), convolutional neural networks (CNN), probabilistic neural networks (PNN), and fuzzy logic.

As noted by Kocur et al. (1996), the objective of electronic breast tumor identification is to decrease the number of false positives (FPs) and to achieve high accuracy for identifying tumors. In breast-growth screenings, incorrect identification of cancerous tumors is more common in breast disease patients [14]. The detecting capacity of the algorithm relies on the number of FPs in an image. Subsequently physicians must devote extra time and care to evaluate the dangerous areas of the images which is marked as FP. The image segmentation is done after removing the edges from the mammographic images. This method helps to find the region of interest (RoI) from the breast images by finding the difference between in edges between normal and abnormal category of breast images. This analysis paves way to study the morphology of the breast tissues for normal and abnormal categories of the breast images [25]. It is not possible to identify the breast mass which represents the repeated accumulation of parenchyma tissues.

Devakumari et al. (2018) proposed a method to detect breast cancer early by using image processing algorithms and data mining methods [26]. Abdullah-Al Nahid et al. (2017) used image-based methods for feature extraction from mammograms and CNN, random forest (RF), support vector machine (SVM), and Baysien classifiers for breast cancer detection [27]. Monica et al. (2016) proposed a feedforward neural network trained with a BPA to detect breast cancer in mammograms. A classification accuracy of nearly 89% was achieved [28].

Vaidehia et al. (2015) developed a method to analyze mammograms using gray-level cooccurrence matrix (GLCM) texture features along with contrast, correlation, energy, and homogeneity to detect breast cancer accurately using a BPA classifier. An accuracy of 93.75% was achieved [29]. Nadeem Tariq (2017) discussed a CAD system for radiologists to increase diagnosis accuracy. In the proposed system, texture features from mammograms were calculated using the GLCM to determine the features which have large impact on diagnosis using artificial neural network (ANN) for training and classification [30].

Karthikeyan et al. (2013) noted that early detection of breast cancer decreases mortality rates and improves survival of women over 40. To improve the sensitivity of the mammogram CAD, mathematical aspects were introduced [31]. Hamido Fujita et al. (2017) developed an image processing-based technique to analyze mammograms by using independent component analysis (ICA). An accuracy of nearly 88.60% was achieved [32]. Raghavendra et al. (2016) proposed an image-based method that uses locality sensitive discriminant analysis (LSDA), decision tree (DT), linear discriminant analysis (LDA), and quadratic discriminant analysis (QDA) with k-nearest neighbor (kNN), naïve Bayes classifier (NBC), PNN, SVM, and fuzzy sugeno classifiers (FSC) to identify normal, benign, and malignant tumors in the breast. The accuracy, sensitivity, and specificity values were 98.69%, 99.34%, and 98.26%, respectively [33].

In 2016, Ayman El-Baz et al. reported that image segmentation with quantitative reasoning is helpful in diagnosis of breast cancer [34]. Indrajeet Kumar et al. (2015) reported on CAD systems that use segmentation approaches to extract the RoI and thereafter the classification was performed using Bidirectional Radial Basis function (BI-RADS) with extracted texture features [35]. Bhusri et al. suggested the use of morphological features like area, perimeter, convex area, diameter, major axis and minor axis length, extent, eccentricity, Euler number, solidity, and orientation can be compared with texture features like GLCM, and Edges with SVM-based classification to identify breast lesions [36].

Jitendra Virumani et al., in 2016, used the database MIAS consisting of 322 samples with 106 fatty and 216 dense images. Features like SD, entropy, kurtosis, skewness, and texture were used to identify breast cancer using SVM and PNN. The classification accuracy was 92.5%. Similarly principal component analysis was used to achieve an accuracy of 94.4% [37]. Soraya Cheriguene et al., in 2016, noted that a CAD approach would improve the diagnosis of breast cancer using a novel classification approach [38]. Similarly in 2016, Nawel Zemmal et al. suggested that a CAD-based approach could improve the diagnosis of breast cancer by using semisupervised learning (SSL) and semisupervised support vector machine (S3VM) to identify breast cancer [38]. Aindrila Bhattacherjee et al., in 2016, proposed a BPA with nine features capable of diagnosing breast cancer with an accuracy of 99.27% [39].

As can be seen, many researchers have extracted a variety of features like mode, mean, texture features, kurtosis, eccentricity, entropy, and major and minor axis length with classifiers like SSM, S3VN, Bayes, DT, PCA, LDA, FSC, and PNN. The maximum accuracy obtained was 99.27%. To increase the classification accuracy to 100% a combination of features like mean, mode, SD, median, area, MAD and texture features can be used with adaptive neurofuzzy structure (ANFIS) classification.

Interpretation and analysis of medical images is one of the most difficult processes in advanced applications of pattern recognition and computer vision. In digital mammography, the main difficulty is discriminating between benign and malignant tumors. In the case of X-ray mammography, the rate of failure is from 4% to 34%.

The ANFIS-based classification uses the feature set area, mean, standard deviation, mode, and median combined with a texture feature for breast cancer diagnosis at an early stage. This method also avoids unnecessary capturing of mammograms and other invasive diagnosis methods harmful to the human body.

Here, we propose an image processing algorithm that can find abnormalities in the breast at an early stage from mammograms. The main goal of this work is to improve the investigative accurateness of image processing for optimal grouping among malignant and benign anomalies at various stages from digital mammograms by reducing the rate of missed cancers. The main aim of this work is to provide an algorithm for high-performance classifier for characterizing breast nodules as benign or malignant. This is also done to detect various stages of malignancy and also to relive the patients from the pain caused due to repeated mammogram testing done to ensure the presence of microcalcifications in the breast.

5.3 Methodology

The immeasurable development of an explicit cluster of cells in a human body's part is referred as CANCER. These tumor cells are known as malicious tumors if

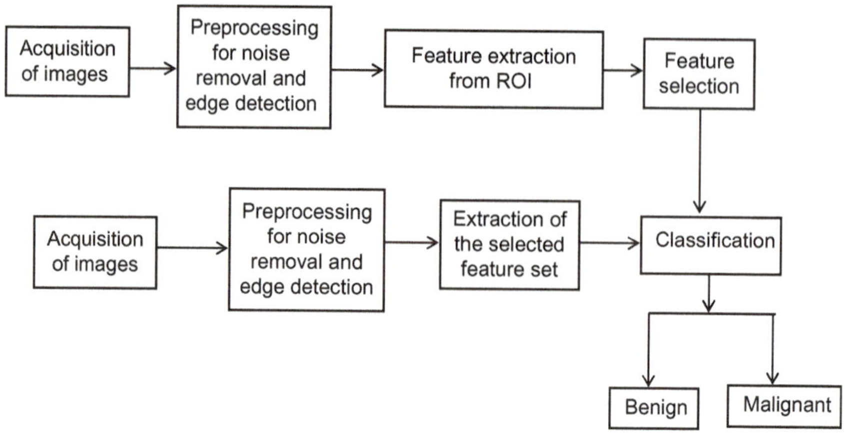

FIGURE 5.5

Block diagram of the proposed work.

they undergo uncontrolled division. Breast cells were the foundation for establishing malignant tumor, recognized as breast cancer. Screening of breast cancer mostly precedes an anatomic methodology over X-ray Mammography which necessitates the breast tumor to be advanced to a point where it is significantly thicker than healthier tissue.

A block diagram of the proposed work is given in Fig. 5.5. To improve accuracy, this system uses machine vision algorithms. The MRI images were collected from the MIAS database.

The important steps include the collection of database pertaining to normal breast, benign and malignant tumors in breast. The database is taken from MIAS. It contains nearly 500 images in the above said three categories. These images were preprocessed for noise removal. Edge detection was also done. Thereafter features like SD, mean, mode, median, MAD from wavelet transform and texture features were extracted. The classification was done using ANFIS and its performance was also compared with an ANN trained using a BPA.

5.3.1 Two-way classification and feature extraction technique

Two way classification denotes two algorithms namely K-means algorithm and Support Vector Machine (SVM) used for diagnosing the cancerous region. Mammogram result is taken as an input and given to preprocessing phase for filtering the data as it may be corrupted with noise. Preprocessing refers to processing the image before feature extraction to make it free from noise. This method will help the classifiers to produce an error free output.

The filter considers an image of area 3×3, 5×5, 7×7, etc., as an element array resulting from an array of pixel values. The median value of an array is

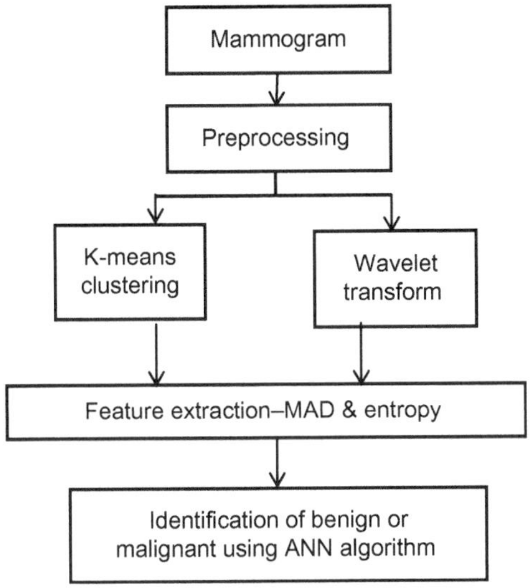

FIGURE 5.6

Two-way classification and feature extraction architecture.

calculated and resulted by ordering element array. A famous sorting technique, called bubble sort, is used in this element array sorting in an ascending order, which returns a median value from the middle elements of the sorted array. The set, the median values of the array elements calculated for all the pixels, were resulted to an output image array. The complete image array is arrived by repeating the median filter process. The following algorithms were used to do preprocessing for noise removal and edge detection as well as training and testing of features.

1. K-means algorithm
2. Wavelet transform

After preprocessing, first K-means algorithm is used and clusters are formed. Once again wavelet transform is applied to the same original image. Then the clustered data and the feature extracted from wavelet transform is given to the SVM algorithm for classification as shown in Fig. 5.6.

5.3.2 K-means algorithm

Mammogram result is taken as an input data for the preprocessing stage. By using a movement of isolating methodologies, racket reachable in the photo can be filtered through. The photo dull regard is reduced when the high pass channel takes after with the movements and sending those movements to a low-pass channel.

So the photo is smoothed and the sharp edges are expelled, while executing with the low-pass channel. This method is reiterated to get the whole image cluster.

A set $D = \{xi \mid i = 1, \ldots, N\}$, where xi implies the ith data point

Set of d-dimensional vectors. The strategy starts with defining k centers and approaching towards formation of centroids. The basic value of 'k' is taken and based on this value the entire data is divided into 'k' subsets for 'k' times until the 'k' value is finalized. Reiterate this computation methodology till the convergence is reached.

Step 1: Assigning Data from set D

Every data point from set D is consigned to its closest centroid, with ties self-self-assuredly broken. Data allocating happen.

Step 2: "connotes" Relocation

Each pack operator data is supplanted to within (mean) of the extensive number of data centers allotted to it. The substitution is to the goals (weighted mean) of the data packages happening if the data centers accomplished the probability measure (weights). Euclidean division is the default measure of closeness, in the midst of this circumstance, nonnegative cost work applies reliably in Eq. (5.1).

$$\sum (\text{argmin} \mid\mid Xi - Cj \mid\mid 2) \tag{5.1}$$

There are reliably K clusters. There is reliably no short of what one thing in each gathering. The groups are nondynamic and they don't cover the Region of Interest. Every person from a gathering is closer to its cluster than some other gathering since closeness does not by and large incorporate "within" of gatherings. K-suggests gathering particularly while using heuristics, for instance, Lloyd's computation seems to be genuinely easy to finish and relate even on broad enlightening accumulations.

Fig. 5.7 shows the flow diagram of the k-means algorithm. Fig. 5.7 shows the flowchart of the k-means count.

This is a basic technique which helps to categorize the data into k-groups. The primary thought is to characterize k centroids, one for each bunch. These centroids are to be put astutely as a result of various parts sources of diverse outcome.

Clustering is a technique used to isolate an arrangement of information into a particular number of gatherings. It is one of the most prominent technique is k-implies grouping. If k implies grouping, it means gathering the information into 'k' groups [11,12]. It groups a given arrangement of information into k number of disjoint bunches. K-implies calculation includes two separate stages. In the principal stage it figures out the k centroid and in the second stage it indicates the clusters which are close to the centroids using minimum distance as the objective. There are diverse techniques used to characterize the separation of the closest centroid; one of the most common is Euclidean separation, which we use here.

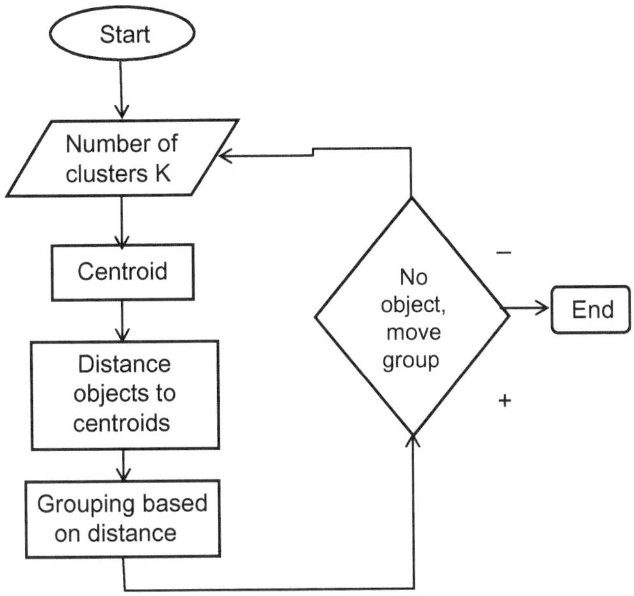

FIGURE 5.7

Flowchart of k-means algorithm.

Once the grouping is done it recalculates the centroid from each group using Euclidean distance and considers the one which has minimum Euclidean distance. Each group of clustered data is characterized by its centroid. The centroid for each bunch is the indication for which the whole of separations from every one of the articles in that group is limited.

So k-means is an iterative calculation that limits the whole of separations from each protest its bunch centroid, over all groups. Give us a chance to consider a image with determination of $x \times y$ and the image must be bunched into k number of groups. Let $p(x, y)$ be information about the pixels to be group and ck be the bunch focuses.

Calculation for k-means clustering:

1. Step 1: Give the number of cluster worth as k.
2. Step 2: Arbitrarily pick the k group focuses.
3. Step 3:Ascertain mean or focal point of the group.
4. Step 4: Ascertain the separation between every pixel to every group focus.
5. Step 5: On the off chance that the separation is close to the inside then move to that bunch.
6. Step 6: Generally move to next bunch.
7. Step 7: Re-appraise the inside.
8. Step 8: Rehash the procedure until the inside doesn't move.

Computational many-sided quality is another term we have to consider while planning the k-implies grouping. It depends on the quantity of information components, number of groups, and number of cycles. After division is done, the fragmented image may even exhibit undesirable areas or commotion. So to make the image better quality, the middle channel is connected to the sectioned image. So to make the image better quality, the middle channel is connected to the sectioned image and also the *n* x *n* section of the image may be utilized for this purpose.

The image preprocessing helps to produce noise free images by identifying the appropriate edges of the images by using edge detection technique. This kind of smoothening operation removes noise.

There are two types of channels: linear and non-linear. The linear channels obey Superposition principle whereas the non-linear channels do not satisfy the Superposition theorem.

The fundamental impediment of direct channel is that the clamor is not evacuated totally and some measure of helpful data is lost. There are also other types of channels like subsidiary channel, Laplacian channel, etc. Controlling the coefficients of the filter helps to remove the noise effectively.

5.3.3 Adaptive neurofuzzy structure

This technique combines the knowledge representation of a fuzzy interference system (FIS) with the learning power of ANNs by combining the features of fuzzy systems and neural networks. ANFIS constructs a FIS by using a given input and output dataset. A BPA is used to adjust the membership function parameters of ANFIS or a combination of a BPA with a least squares type of method is used. ANFIS combines NN and fuzzy logic, which are two machine learning techniques. ANFIS works by applying the learning methods of NN to tune the parameters of a FIS.

The ANFIS architecture is illustrated in Fig. 5.8, in which a circle represents a fixed node and a square represents an adaptive node. The essential parts of a fuzzy inference system are found in five layers. All the nodes are adaptive in layer 1 and the outputs of this layer are the fuzzy membership grades of the inputs. In layer 2, all the nodes are fixed nodes. This layer uses AND operator which is a fuzzy operator to fuzzify the inputs. This operator in the figure is labeled with $\sum$, representing a multiplier. In layer 3, all nodes are fixed nodes that play a normalization role. This normalization layer uses the current values of all fuzzy variables to calculate the value of the mass firing of all rules. In layer 4, all the nodes are adaptive and the output of each node is the product of the normalized strength. In layer 5, only one node labeled $\sum$ performs the summation of all incoming signals. The ANFIS refines fuzzy IF-THEN rules to describe the behavior of the system. The learning algorithm of ANFIS is a combination of gradient descent and least squares methods.

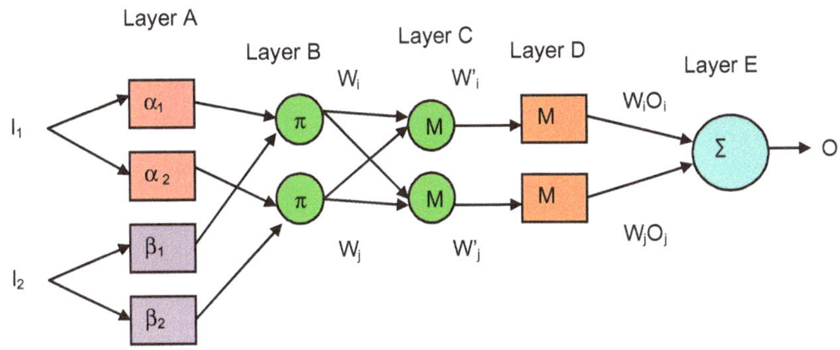

FIGURE 5.8

Architecture of ANFIS.

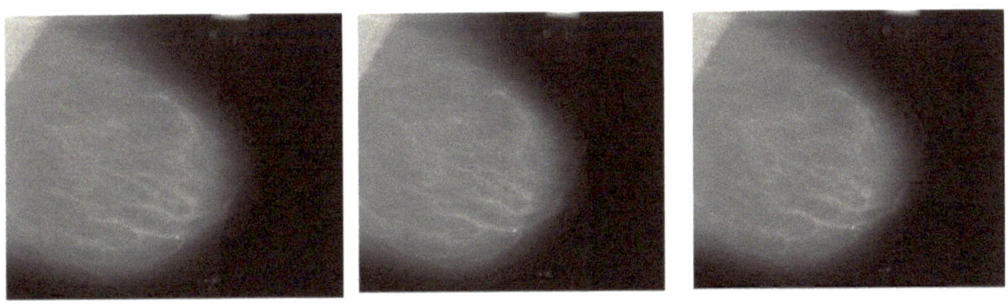

FIGURE 5.9

Mammogram image without tumor—samples of normal image.

5.4 Results for diagnosis of microcalcification in breast

5.4.1 Image acquisition

An opinion from the radiologist is obtained and used as the Observer decision before implementing the image processing algorithms (machine vision algorithms) for identification. In general 75% of the data is used for training, 15% for testing, and the remaining 10% for validation. A total of 500 images of the breast were gathered from MIAS out of which 366 samples were used for training, 75 images for testing, and the remaining 59 images for validation. Various image processing algorithms and AI classifiers were modeled using MATLAB software. The samples of digital mammograms were given as the typical data for the proposed work.

The mammogram image without a tumor is shown in Fig. 5.9 and the presence of a solid tumor (malignant and benign) is shown in Fig. 5.10A and B, respectively.

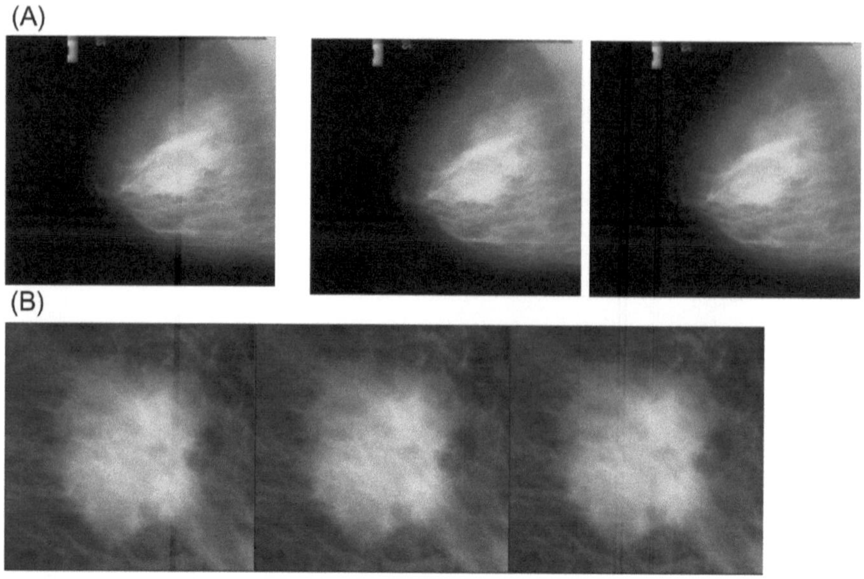

FIGURE 5.10

(A) Samples of mammogram images with malignant tumor. (B) Samples of mammogram images with benign tumor.

5.4.2 Preprocessing

The preprocessing incorporates separation and edge detection. The separation of edges produces noise free images and can be used for further investigation. Edge detection is done to locate the RoI from which features are extracted. The following algorithms were used to do preprocessing for noise removal and edge detection and training and testing of features.

5.4.3 Edge detection

The edge detection algorithm was used to separate foreground from background. Take the gradient of image and find the difference. If the difference is zero then the regions are similar. If there is a difference a boundary is detected.

The edge detection for a normal image is shown in Fig. 5.11. The edge-detected image for an abnormal image is shown in Fig. 5.12. From Fig. 5.12, changes in the image brightness of the abnormal image can be seen. The discontinuities which has significant warped ends depends on the image intensity. The segmented abnormal image is shown in Fig. 5.13, and the segmented normal image is shown in Fig. 5.14. The abnormal image is made into a set of segments that collectively cover the entire image. Fig. 5.14 shows the location of the tumor region and its boundaries.

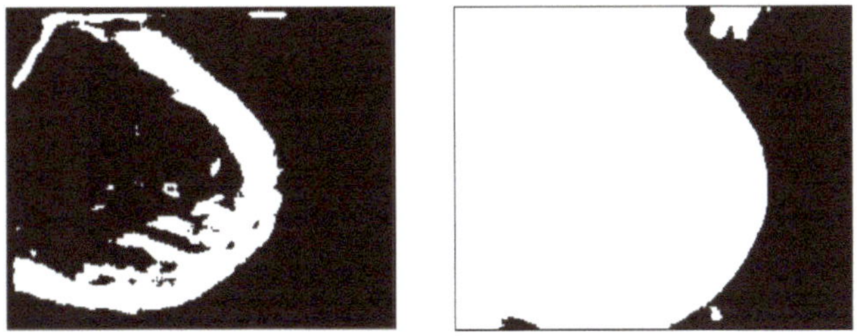

FIGURE 5.11

Edge detection for normal image of the breast.

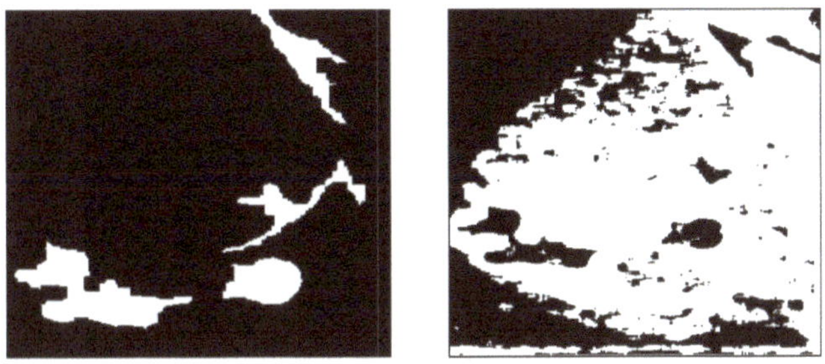

FIGURE 5.12

Edge detection for abnormal image of the breast.

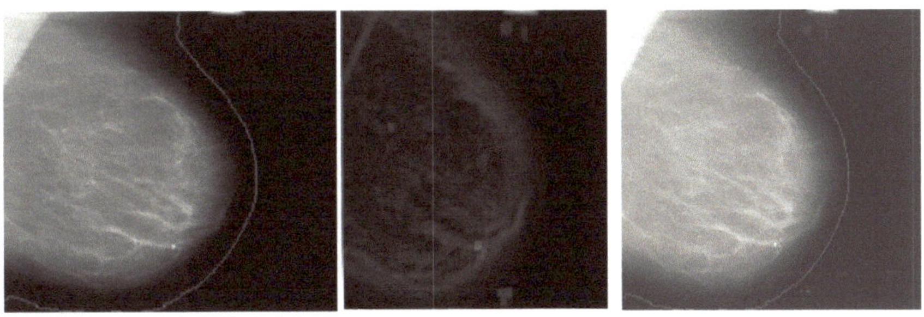

FIGURE 5.13

Segmented output for normal image of the breast.

5.4.4 Feature extraction

To implement testing and training using ANN and ANFIS the features needs to be separated into three sets, i.e., the training set and the testing/validation set. The features extracted from the k-means algorithm are median, area, mean, standard deviation, and mode. The features extracted from the wavelet transform are the MAD and approximation values of the wavelet coefficients.

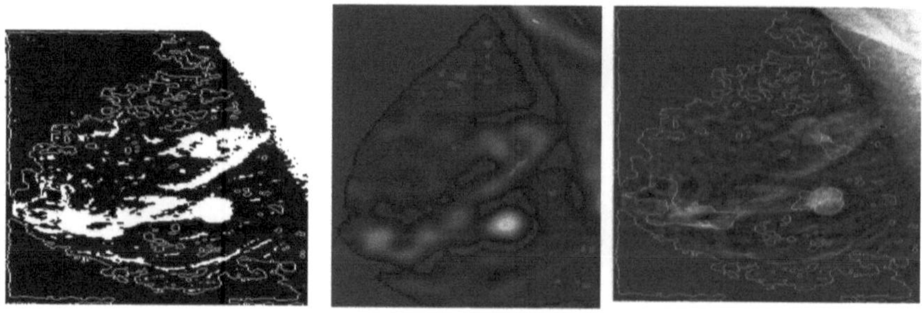

FIGURE 5.14

Segmented output for abnormal image of the breast.

Table 5.1 Features extracted from normal breast images.

Sl. No.	Area (Sq. pixels)	Mean (pixels)	Std. Dev. (pixels)	Mode (pixels)	Median (pixels)	MAD (no units)	Entropy of the texture
1.	232487	51.102	28.908	23	40	101.5	3.4887
2.	270336	66.095	39.951	22	41	101.8	3.4584
3.	270336	49.631	26.238	21	48	104.4	3.4919
4.	352256	63.419	35.706	20	44	105.3	3.4100
5.	352256	44.944	22.102	24	42	106.7	3.5650
6.	352256	64.591	36.62	25	40	108.2	3.7659
7.	352256	43.843	30.295	27	47	109.2	3.7722

Table 5.1 lists the various features extracted from the normal images. Table 5.2 lists the same features extracted from the abnormal images. From Table 5.1 and Table 5.2, it can be seen that the area of the tumor region of the abnormal image is larger than the area obtained in the normal images and MAD value is also higher than the normal images.

The feature represents the basic characteristic information in the image which gets repeated itself in various directions within the image. The various features include area, mean standard deviation mode and median, area, mean, standard deviation, mode (extracted using Image J) combined with MAD (from wavelet transform) for classification by ANN. These features are primarily normalized that is then used for categorization of benign or malignant tumors. The MAD value is also higher in the case of abnormal images as compared to the normal images of the breast.

5.4.5 Performance evaluation

The performance evaluation is done using various parameters like true positive (TP), FP, true negative (TN), and false negative (FN). TP (here) is defined as the

Table 5.2 Features extracted from abnormal breast images.

Sl. No.	Area (Sq. pixels)	Mean (pixels)	Std. Dev. (pixels)	Mode (pixels)	Median (pixels)	MAD (no units)	Entropy of the texture
1.	88209	82.768	40.647	22.12	74	211.5	4.6803
2.	86757	71.032	34.648	24.38	70	221.3	4.7297
3.	88112	79.72	34.834	25.12	73	232.2	4.7469
4.	86632	72.625	33.881	26.22	71	236.4	4.8076
5.	88088	85.364	42.912	27.61	73	240.1	5.2222
6.	88610	81.262	40.475	32.7	72	242.8	5.0071
7.	88309	49.958	24.039	34.61	47	248.2	5.0371

Table 5.3 Confusion matrix.

Confusion matrix	Positive	Negative
Positive	TP	FP
Negative	FN	TN

images correctly identified as benign or malignant. TN is defined as malignant images correctly identified. FN also called as type II error is defined as benign or malignant images incorrectly identified.

Similarly FP is also called a type I error and is defined as benign incorrectly identified. The classification performance of the intelligent classifiers is evaluated using precision and recall. Precision is defined as the ratio of number of all the images retrieved to the number of all the relevant images retrieved. Recall is defined as the ratio of the number of all the relevant images retrieved to the number of all the images in each group. The execution of combined classifier is incomparable with the standalone classifiers because the combined classifiers may reduce the mis-classifications in the test data.

Execution of a paired classifier can be best depicted as far as its affectability and specificity by evaluating its execution of FP and FN cases.

The confusion matrix in Table 5.3 shows the relationship between different performance indices for classification. For evaluating the performance of the image based classification of malignant and benign tumors from digital mammograms, the four performance indices (TP, TN, FP, and FN) in Table 5.4 are used and the same is compared with the l one determined by a biopsy test. Using these four performance measures, relative measurements for binary classification can be calculated.

Recall is also referred to as the TP rate or sensitivity, and precision is also referred to as the positive predictive value; other related measures used in classification include TN rate and accuracy. The TN rate is also called specificity.

Table 5.4 Classification performance measures.

Performance measures	Definition
TP	Tumor marked as malignant by a biopsy, which is also classified as malignant by the learning machine.
FP	Tumor marked as malignant by a biopsy, which is classified as benign by the learning machine.
TN	Tumor marked as benign by a biopsy, which is also classified as benign by the learning machine.
FN	Tumor marked as benign by a biopsy, which is classified as malignant by the learning machine.

The database includes the mammogram images from the MIAS and clinical images from medical practitioners. The first part includes database collection that includes normal and abnormal breast images. These images are preprocessed as an initial step. The second part of this research contracts with feature extraction and feature selection. The different features are extracted by using k-means and wavelet transform. In the third part feature selection is done to check if the feature set improves the performance of the classifiers after thorough analysis. The selected features are used as inputs for the ANFIS. The performance is evaluated using certain metrics like TP, Precision and Recall which are suitable for biomedical image processing.

During this phase the features which enhance the performance of the classifier are exclusively selected. This analysis is found to be beneficial in the biomedical field so that the cancer can be detected at the early stage. This method also avoids unnecessary capturing of mammograms and other invasive diagnosis methods which is harmful for the human body. Finally validation of the proposed algorithms is also done as a part of this investigation. This substantiates that the proposed image processing and artificial intelligence (AI) methods for diagnosis of breast cancer serve the purpose to the maximum possible extent to the medical practioners.

The results for training, testing, and validation of the ANFIS are shown in Figs. 5.15–5.17, respectively. This ANFIS classifier shows robust performance as indicated in Table 5.5.

5.5 Discussions

The efficiency of the proposed strategy is controlled based on the degree to which the noise is removed using preprocessing methods and also the image gets segmented to extract the RoI. The TP rate and the number of FP rate at various maximum limits of the images are used to evaluate the algorithm's performance.

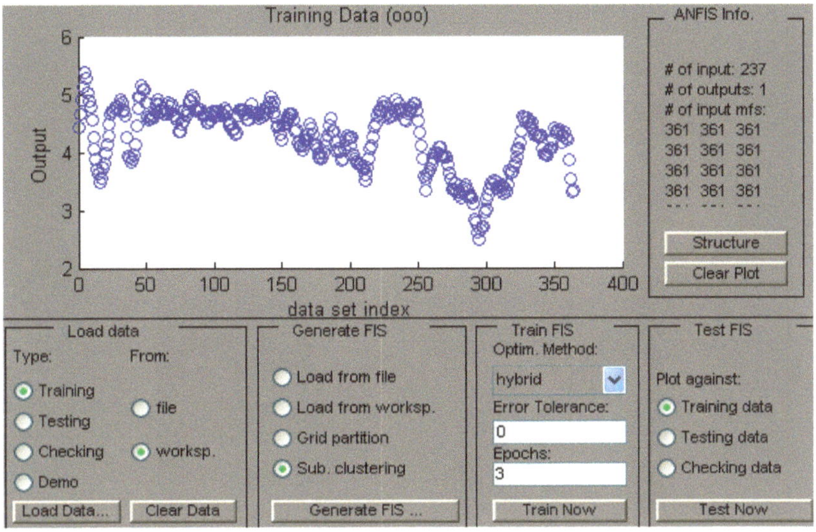

FIGURE 5.15

ANFIS-based training for early detection of breast cancer.

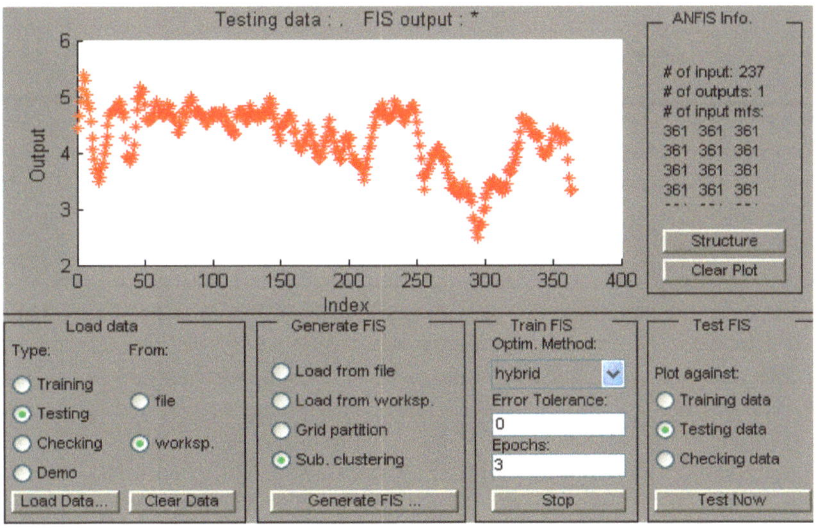

FIGURE 5.16

ANFIS-based testing for early detection of breast cancer.

A comparison of the existing and proposed algorithms is given in Table 5.5. Table 5.6, denotes the comparative results for testing and validation. These results are achieved by using the feature set (area, mean, standard deviation, and mode

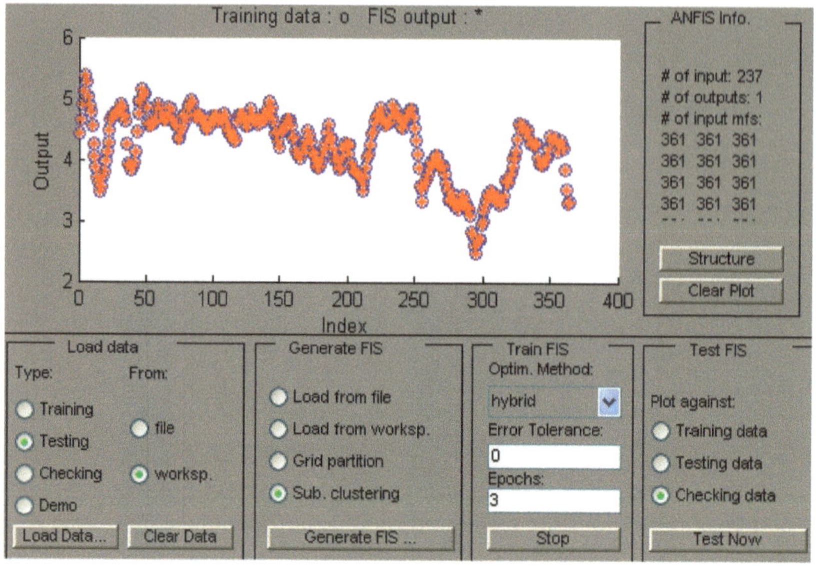

FIGURE 5.17

Validation of ANFIS structure for early detection of breast cancer.

and median from k-means algorithm combined with MAD from wavelet transform) extracted and classified using ANN algorithms.

During the clustering process, the precision and recall values for normal and abnormal images from the k-means algorithms are 0.83 and 0.89, respectively. By using the MAD feature from the wavelet transform the precision and recall values for normal and abnormal images are 0.83 and 0.87, respectively, by the ANN classifier trained with the BPA. If the base feature set is combined with the selected features from the k-means algorithm and wavelet transform with ANFIS classifier the precision and recall values are 0.98 and 0.99. This shows that the performance can be increased by the feature fusion technique. The features from k-means and wavelet transform were combined along with texture features and used as inputs for training the ANFIS. In this case, the precision and recall values are 0.98 and 0.99, respectively. For the remaining methods like ICA with cooccurrence features, LSDA with FSM, PNN, SVM, DT, NBC, and k-NN, the range of maximum and minimum values for precision and recall are 0.7–0.9 and 0.8–0.98 for the abnormal category. For the normal category the min-max range of precision and recall is 0.7–0.9 and 0.8–0.97, respectively.

The performance measurement for optimal classification indicates that the values of recall and precision should be equal to 1. Hence, the ANFIS with texture features (Table 5.6) shows optimal behavior for the classifier for finding normal, benign, or malignant tumor images and their associated values. The features are obtained by both k-means algorithm and wavelet transform, and texture

Table 5.5 Comparison of existing and proposed algorithms.

	Performance measures for classification by various algorithms						
		Abnormal					
		Benign		Malignant		Normal	
Sl. No.	**Type of algorithm**	**Precision**	**Recall**	**Precision**	**Recall**	**Precision**	**Recall**
1.	K-means algorithm	0.83	0.89	0.83	0.89	0.83	0.89
2.	Wavelet transform	0.83	0.87	0.83	0.87	0.83	0.87
3.	**ANFIS-based algorithms using texture as a feature (proposed method)**	**0.98**	**0.99**	**0.98**	**0.99**	**0.98**	**0.99**
4.	ICA with cooccurrence features	0.72	0.9	0.72	0.9	0.72	0.9
5.	LSDA with FSM	0.9	0.92	0.9	0.92	0.9	0.92
6.	LSDA with PNN	0.88	0.98	0.88	0.96	0.88	0.97
7.	LSDA with SVM	0.89	0.98	0.89	0.95	0.89	0.97
8.	LSDA with DT	0.74	0.88	0.74	0.88	0.74	0.88
9.	LSDA with NBC	0.71	0.86	0.71	0.86	0.71	0.86
10.	LSDA with k-NN	0.7	0.8	0.7	0.8	0.7	0.8

Table 5.6 Performance of ANFIS classifier.

Comparison testing/ validation	Recall			Precision		
	Normal	Benign	Malignant	Normal	Benign	Malignant
Testing results	0.98	0.98	0.98	0.97	0.97	0.97
Validation results	0.98	0.98	0.98	0.97	0.97	0.97

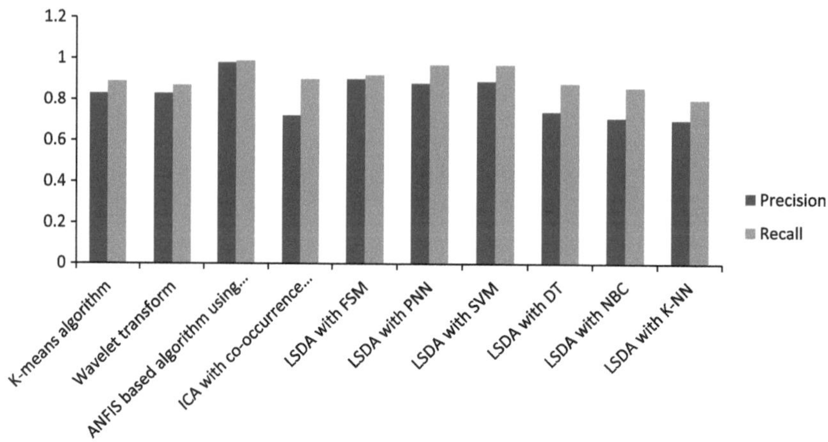

FIGURE 5.18

Performance measures for classification by various algorithms for normal images.

features are used for training and testing the ANFIS and BPA. The same precision and recall for normal and abnormal images by using various other methods are also shown in Table 5.5.

The graphs (precision and recall) for both normal images and abnormal images are shown in Figs. 5.18 and 5.19 for various algorithms. The precision and recall values of both normal and abnormal images are 0.83–0.98 and 0.8–0.99, respectively. To summarize Figs. 5.18 and 5.19 show the comparative performance for various other algorithms and ANFIS structures. The performance measures for ANFIS are shown in Figs. 5.18 and 5.19; the values of precision and recall are closer to 1, which means that the classification is done in a robust manner. The testing/validation of the proposed ANFIS classifier is also done and its performance measures like precision and recall are tabulated in Table 5.6 and depicted in Fig. 5.20, showing that the performance measure values are closer to 1 (0.97 and 0.98) for ANFIS-based classification using the base feature set with the texture features (Fig. 5.20).

Table 5.6 shows that the testing and validated results are 0.97 for precision and 0.98 for recall. Since the performance of the ANFIS classifier is the same for

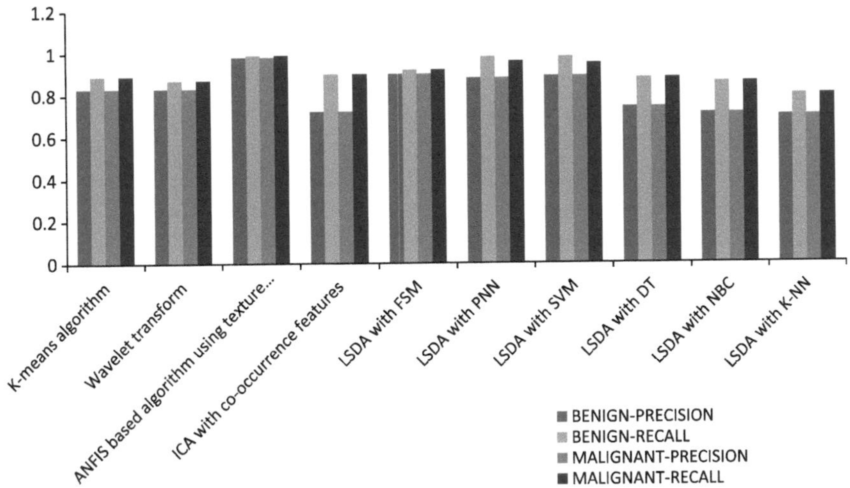

FIGURE 5.19

Performance measures for classification by various algorithms for abnormal images.

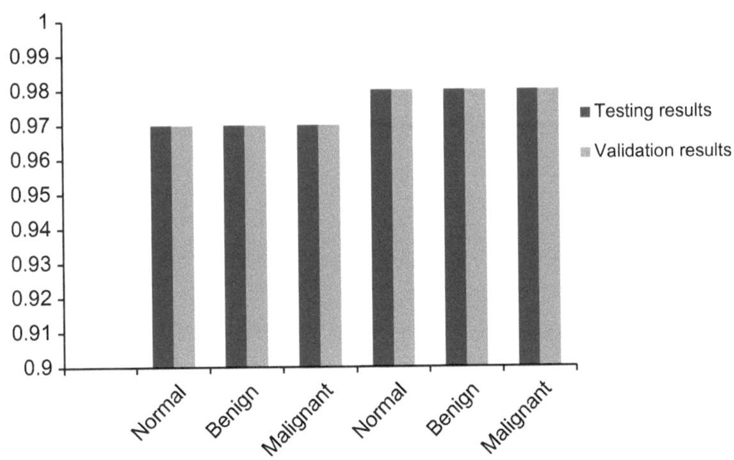

FIGURE 5.20

Testing and validation results for precision and recall for ANFIS using texture as a feature for classification.

testing and validation, this classifier proves to be an efficient one in identification of the tumors. This substantiates that the proposed image processing and AI methods for diagnosis of breast cancer serve the purpose to the maximum possible extent to the medical practioners.

5.6 Conclusion

This chapter discussed the two-way classification architecture and showed how an abstract structure allows us to discover effective classification of breast cancer images. This algorithm is simple and the results are encouraging for a real-time breast cancer diagnosis system. After experimentation it was concluded that the hard threshold does not always give better denoising performance; it depends on which wavelet thresholding algorithm was chosen. Wavelet denoising is a powerful tool for image enhancement.

The performance measurements for k-means, wavelet transform, and other types of algorithms are recall and precision, which were in the range of 0.7–0.99 and 0.8–0.98, respectively, here for training. On the other hand, the performance of the ANFIS classifier is same for training, testing and validation (0.97 and 0.98 and 0.98 and 0.97 for precision and recall) and hence the proposed system's performance can be scaled up and further enhanced by framing new algorithms which is more adaptable to mammograms.

5.7 Future scope

As a part of the future work it can be made possible to diagnose breast cancer at an early stage even for breastfeeding mothers by realizing efficient algorithms along with features like CNN and GLCM due to the variation in the levels of progesterone and estrogen.

References

[1] S. Detounis, Computer aided detection and second reading utility and implementation in a high volume breast clinic, Appl. Radiol. 3 (2004) 8–15.

[2] J.M. Steppe, Feature and model selection in feed forward neural networks, Ph.D dissertation, Air Force Institute of Technology, Air University, United States, 1994.

[3] B.S. Monsees, Evaluation of breast microcalcifications, Radiol. Clin. N Am. Breast Imag. 33 (1995) 1109–1121.

[4] S.A. Feig, M.J. Yaffe, Digital mammography, computer-aided diagnosis and telemammography, Radiol. Clin. N Am. Breast Imag. 33 (1995) 1205–1230.

[5] L.W. Bassett, V.P. Jackson, R. Jahan, Y.S. Fu, R.H. Gold, Diagnosis of Diseases of the Breast, W.B. Saunders Company, 1997.

[6] American Cancer Society, Breast Cancer: Facts and Figures 2003–2004, American Cancer Society, Atlanta, GA, 2003.

[7] W.P. Evans, Breast masses appropriate evaluation, Radiol. Clin. N Am. Breast Imag. 33 (1995) 1085–1108.

[8] S. Buseman, J. Mouchawar, N. Calonge, T. Byers, Mammography screening matters for young women with breast carcinoma, Cancer 97 (2003) 352–358.

[9] G.M. Palmer, C. Zhu, T.M. Breslin, F. Xu, K.W. Gilchrist, N. Ramanujam, Comparison of multiexcitation fluorescence and diffuse reflectance spectroscopy for the diagnosis of breast cancer, IEEE Trans. Biomed. Eng. 50 (2003) 1233–1242.

[10] A.M. Khuzi, R. Besar, W.M.D.W. Zaki, N.N. Ahmad, Identification of masses in digital mammogram using gray level co-occurrence matrices, Biomed. Imag. Interv. J. 5 (2009) 1–13.

[11] R.M. Rangayyan, N.M. El-Faramawy, J.E.L. Desautels, O.A. Alim, Measures of acutance and shape for classification of breast tumors, IEEE Trans. Med. Imag. 16 (1997) 799–810.

[12] D. Guliato, R.M. Rangayyan, J.D. de Carvalho, S.A. Santiago, Spiculation preserving polygonal modeling of contours of breast tumors, in: Proceedings of the 28th Annual International Conference of the IEEE Engineering in Medicine and Biology Society (EMBS), New York, 2006, pp. 2791–2794.

[13] J. Bozek, K. Delac, M. Grgic, Computer-aided detection and diagnosis of breast abnormalities in digital mammography, in: Proceedings of the 50th International Symposium ELMAR, Zadar, 2008, pp. 45–52.

[14] R. Mutihac, A.A. Colavita, A. Cicuttin, A. Cerdeira, Maximum entropy improvement of x-ray digital mammograms, in: N. Karssemeijer, M.A.O. Thijssen, J.H.C.L. Hendriks, L.J.T.O. van Erning (Eds.), Digital Mammography, Kluwer, Dordrecht, 1998, pp. 329–336.

[15] E. Cernadas, L. Gomez, P.G. Rodriguez, A. Casas, R.G. Carrion, J.J. Vidal, Design of un sharp masking filters in the frequency domain: parametrization for breast radiographs, in: K. Doi, M.L. Giger, R.M. Nishikawa, R.A. Schmidt (Eds.), Digital Mammography, Elsevier, Amsterdam, 1996, pp. 463–466.

[16] R.N. Strickland, L.J. Baig, W.J. Dallas, E.A. Krupinski, Wavelet-based image enhancement as an instrument for viewing cad data, in: K. Doi, M.L. Giger, R.M. Nishikawa, R.A. Schmidt (Eds.), Digital Mammography, Elsevier, Amsterdam, 1996, pp. 441–446.

[17] S.R. Aylward, B.M. Hemminger, E.D. Pisano, R.E. Johnston, Mixture modeling for digital mammogram display and analysis, in: N. Karssemeijer, M.A.O. Thijssen, J.H. C.L. Hendriks, L.J.T.O. van Erning (Eds.), Digital Mammography, Kluwer, Dordrecht, 1998, pp. 305–312.

[18] T. Netsch, M. Biel, H.O. Peitgen, Display of high-resolution digital mammograms on CRT monitors, in: N. Karssemeijer, M.A.O. Thijssen, J.H.C.L. Hendriks, L.J.T.O. van Erning (Eds.), Digital Mammography, Kluwer, Dordrecht, 1998, pp. 313–320.

[19] N. Karssemeijer, G.M. te Brake, Detection of stellate distortions in mammograms, IEEE Trans. Med. Imaging 15 (1996) 611–619.

[20] J.W. Byng, J.P. Critten, M.J. Yaffe, Thickness-equalization processing for mammographic images, Radiology 203 (1997) 564–568.

[21] N. Nicolaou, S. Petroudi, J. Georgiou, M. Polycarpou, M. Brady, Towards pectoral muscle removal via Independent Component Analysis, in: Proceedings of 4th IET International Conference on Advances in Medical, Signal and Information Processing, 2008, pp. 1–4.

[22] N. Angayarkanni, D. Kumar, G. Arunachalam, The applications of image processing techniques for detection and classification of cancerous tissue in digital mammogram, J. Pharm. Sci. Res. 8 (10) (2016) 1179–1183.

[23] M.M. Mehdy, P.Y. Ng, I.E.F. Shair, N.I. Md Saleh, Artificial neural networks in image processing for early detection of breast cancer, Comp. Math. Methods Med. 2017 (2017) 15.

[24] A.K. Singh, B. Gupta, A novel approach for breast cancer detection and segmentation in a mammogram, Proc. Comp. Sci. 54 (2015) 676–682.
[25] J. Shell, W.D. Gregory, Efficient cancer detection using multiple neural networks, IEEE J. Trans. Eng. Health Med. 5 (2017) 2800607.
[26] A.A. Nahid, Y. Kong, Involvement of machine learning for breast cancer image classification: a survey, Comp. Math. Methods Med. 2017 (2017) 29.
[27] D. Selvathi, A. Aarthy Poornila, Breast cancer detection in mammogram images using deep learning techniques, Middle East J. Sci. Res. 25 (2) (2017) 417–426.
[28] Monica, Singh Sanjay Kumar, Breast cancer diagnosis using digital image segmentation techniques, Indian J. Sci. Technol. 9 (28) (2016) 617–624.
[29] K. Vaidehia, T.S. Subashini, Automatic characterization of benign and malignant masses in mammography, Proc. Comput. Sci. 46 (2015) 1762–1769.
[30] N. Tariq, Breast cancer detection using artificial neural networks, J. Mol. Biomark. Diagn 9 (1) (2017) 1.
[31] K. Ganesan, U.R. Acharya, C.K. Chua, L.C. Min, K.T. Abraham, K.-H. Ng, Computer aided breast cancer detection using mammogram: a review, IEEE Rev. Biomed. Eng. 6 (2016) 77–98.
[32] H. Fujita, U. Raghavendra, A. Gudigar, V.V. Vadakkepat,U.R. Acharya, International conference on new trends in intelligent software methodology tools and techniques, 2017.
[33] U. Raghavendra, U.R. Acharya, H. Fujita, A. Gudigar, J.-H. Tan, S. Chokkadi, Application of Gabor wavelet and locality sensitive discriminant analysis for automated identification of breast cancer using digitized mammogram images, Appl. Soft Comput. 46 (2016) 151–161.
[34] A. El-Baz, X. Jiang, J. Suri, Biomedical Image Segmentation: Advances and Trends, CRC Press, 2016.
[35] I. Kumar, J. Virmani, H.S. Bhaduria, A review of breast density classification methods, in: IEEE Conference on Computing for Sustainable Global Development, 2015.
[36] S. Bhusri, S. Jain, J. Virmani, Breast lesions classification using the amalagation of morphological and texture features, Int. J. Pharm. BioSci. 7 (2) (2016) 617–624.
[37] J. Virmani, N. Dey, V. Kumar, PCA-PNN and PCA-SVM based CAD systems for Breast Density Classification, Appl. Intel. Opt. Biol. Med. 2016 (2016) 159–180.
[38] S. Cheriguene, N. Azizi, N. Zemmal, N. Dey, H. Djellali, N. Farah, Optimized tumor breast cancer classification using combining random subspace and static classifiers selection paradigms, Appl. Intel. Opt. Biol. Med. (2016) 289–307.
[39] N. Zemmal, N. Azizi, N. Dey, M. Sellami, Adaptive semi supervised support vector machine semi supervised learning with features cooperation for breast cancer classification, J. Med.-Imag. Health Inform. 6 (1) (2016) 53–62.

Further reading

A. Bhatacharjee, S. Roy, S. Paul, P. Roy, N. Kumar, N. Dey, Classification approach for breast cancer detection using back propagation neural network: a study, in: Biomedical Image Analysis and Mining Techniques for Improved Health Outcomes, IGI Global, 2016.

CHAPTER

Work environment and healthcare: a biometeorological approach based on wearables

6

Pablo Fdez-Arroyabe[1], Diego Soliño Fernández[2] and Jon Bilbatua Andrés[3]

[1]*Geography Department, University of Cantabria, Cantabria, Spain*

[2]*Lead Business Development at WearHealth UG, WearHealth, Spain*

[3]*Markontzaga Health Center, Osakidetza-Basque Health System, Sestao, Spain*

Chapter Outline

6.1 Introduction

Recent technological advances such as mobile devices, smart algorithms, wearable technology, and the existence of relevant health databases have unprecedented potential for improving health and wellbeing by transforming clinical practice and self-management [1,2]. Among those innovations, the Internet of Things (IoT) refers to roughly 50 billion smart objects that can connect to the Internet and use storage infrastructures in the cloud [3]. When these devices are incorporated into clothing items or accessories, they are known as wearable

Wearable and Implantable Medical Devices. DOI: https://doi.org/10.1016/B978-0-12-815369-7.00006-9

devices or wearables. Wearable technology consists in biosensors that allow continuous or frequent measurement of health parameters such as body temperature, pulse rate, blood pressure, electrocardiogram, blood sugar, blood oxygen, and sleep patterns in a very convenient manner [2,4].

The number of connected wearable devices worldwide is expected to reach 601 million in 2020 [5]. There are at least 500 different health-related wearables, and the number is likely to increase in the future [6]. In addition to the increasing usage of wearable devices, advances in the reliability and accuracy of wearable devices and its capacity to record and share information online in real time, combined with new algorithms for big data mining and analytics, create new opportunities in the healthcare sector. These technological advances can be used to foster personalized care, focus on prevention rather than disease treatment, routinely monitor personalized physiological measurements, improve self-management, identify alterations in health conditions, and create positive long-term behavioral changes towards healthy lifestyles [2,3,7].

In contrast, there is limited scientific evidence about the use of wearable technology in healthcare and its impact [4]. On the one hand, some studies point out that wearable sensors could cause overdiagnosis and could not have a significant impact on healthcare costs [4,8] or healthcare utilization. On the other hand, some studies show that self-tracking can empower individuals and improve health and wellness outcomes [9,10].

With regards to the adoption of wearable technology, it is important to mention that the quantified-self movement (QSM), which can be defined as "any individual engaged in the self-tracking of any kind of biological, physical, behavioral, or environmental information [...] with a proactive stance toward obtaining information and acting on it," is growing fast [11]. For instance, it is estimated that 60% of US adults track their weight, diet, or physical activity and at least 33% monitor other health-related factors such as blood sugar, headaches, blood pressure, or sleep patterns [12,13]. Furthermore, wearable-related health data at an aggregate level is being adopted by researchers to crowd source public health research in diverse ways [11].

Nevertheless, regarding self-tracking, there are practical and mindset barriers to widespread adoption (i.e., nonpassive data collection, no financial incentives, usability, and comfort). Respecting usability and comfort, wearable devices have experienced major improvements, and thus are more likely to be adopted by the general public. With regards to financial incentives, employers are incentivized to reduce healthcare costs and foster health promotion [11] in the workplace. Finally, regarding mindset, there might be deep cultural, sociological, and philosophical issues but those could be overcome with adequate marketing.

Wearables could also help to achieve United Nations Millennium Development Goals (MDG) related to climate mitigation and its impact on human health. Technological development and the creation of digital devices is essential to keep up with the new scientific challenges brought-up by climate change, such as the impact of weather and climate on living organisms, particularly human beings. Today, many wearables register atmospheric data such as temperature, air humidity, and sun radiation inside and outside building, which represents an excellent opportunity to analyze the impact of the thermal environment on human health in virtually every location (i.e., at home, the workplace, or school).

Due to global warming, climate change has become the biggest global threat for living organisms on Earth. According to the IPCC Special Report on the impacts of global warming of 1.5°C "it is expected that further increases in the number of warm days/nights and decreases in the number of cold days/nights, and an increase in the overall temperature of hot and cold extremes."

To help solve this problem, the World Meteorological Organization (WMO) and the World Health Organization (WHO) have established the first joint initiative to develop the Global Framework for Climate Services (GFCS) with four main research lines: food, water, energy, and health. The GFCS refers to its vision as "to enable better management of the risk of climate variability and change and adaptation to climate change, through the development and incorporation of science-based climate information and prediction into planning, policy and practice on the global regional and national scale" (Fig. 6.1).

The implementation of the GFCS is based on different components, out of which "Observation and Monitoring" is the preliminary one. Seasonal and subseasonal forecasting is based on long periods of data collection (years or decades) to elaborate predictions and projections based on models and emissions scenarios.

At a meteorological scale, the integration of heterogeneous data (health and climatic data) is a must [14]. However, the delivery of weather-related health services, such as the ones outlined in this chapter, represents an important challenge, particularly if done daily.

There are already some services related to heat waves, allergens or air pollution, but there are still many opportunities for innovative companies to develop health-related customized services based on day-to-day weather forecasting [15] at a local scale and in multiple environments.

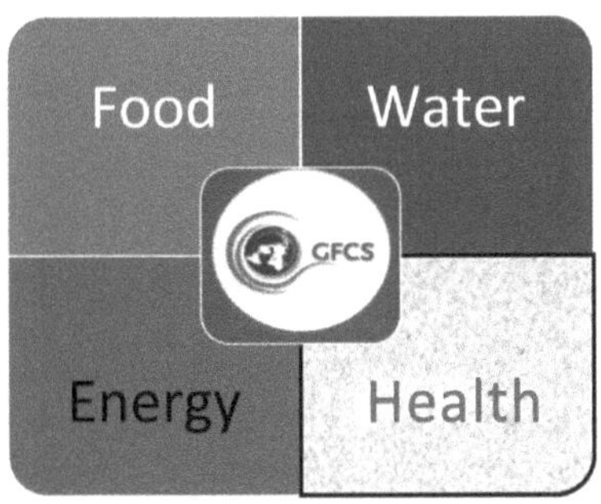

FIGURE 6.1

Based on the Global Framework for Climate Services (GFCS).

Courtesy of WearHealth. https://www.wmo.int/gfcs/about-gfcs.

Wearables devices can contribute to this new wave of innovative services by monitoring human physiology in real time. The collection and integration of heterogeneous data (meteorological, environmental, and physiological) are now achievable at a much shorter temporal scale (hours, days), which makes possible to control human health and deliver weather-related health services.

The biometeorological approach operates at the weather scale [16] and combines the usage of health-related wearables and meteorological sensors. Data from these sensors could be used in multiple applications and disciplines such as urban planning; safety and productivity at work; energy balance in buildings and sustainable development; public health epidemics spreading and warning systems development. Authorities already consider heat stress and thermal comfort as important occupational health and safety risk factor across various sectors including agriculture, fishing, construction, mining, and manufacturing. Even though, exposure to extreme heat represents a major risk to worker health and safety, traditional meteorological weather stations cannot provide accurate data about the closest thermal environment of a worker in real time. Fortunately, this could be solved with the usage of wearable and meteorological sensors.

All in all, given the current state of scientific evidence concerning wearables, it is not clear if wearable technology and wearable data-driven analytics will be successfully implemented because, among other things, the penetration of such technology could take years or even decades [2]. This unpredictability is partly related to the fact that external forces such as industry players, funding, public policy, technology, customers [2,17] and accountability can foster or kill wearable-based innovations.

The topics presented in this chapter were obtained from multiple sources including an in-depth literature review in relevant databases such as Scopus and Web of Knowledge, textbooks and manuals. To create an original while didactic chapter, some sections include comments and examples from the personal experience and expertise of the authors who represent the three main sectors (academia, medical and industry) (Fig. 6.2) needed to transform scientific knowledge into climate-related health services based on digital technologies.

6.2 Biometeorological framework for the use of wearables

6.2.1 Wearable devices and wearable sensors

The use of wearable and implantable devices in various daily situations is a complex process that involves different scientific disciplines and stakeholders. To clarify this process, Fig. 6.3 describes a three-step framework to define, test and validate wearable-driven solutions in various scenarios.

As in many other processes, the first step is to define the goals to be achieved. The implementation of wearable devices is no different and goes hand in hand with the problems that could be solved with their insights based on real time data.

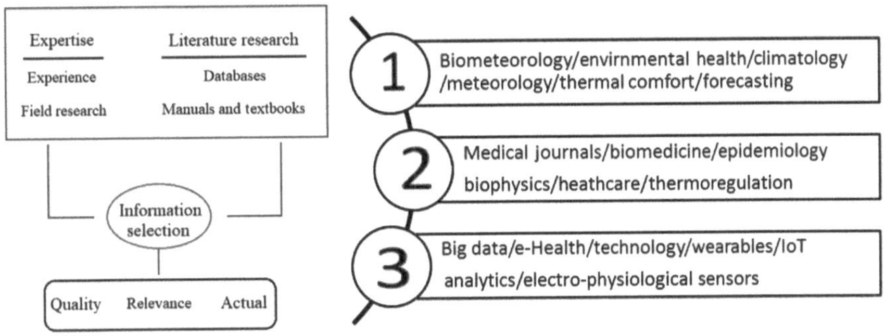

FIGURE 6.2

Sources of information and author's expertise. Courtesy of WearHealth.

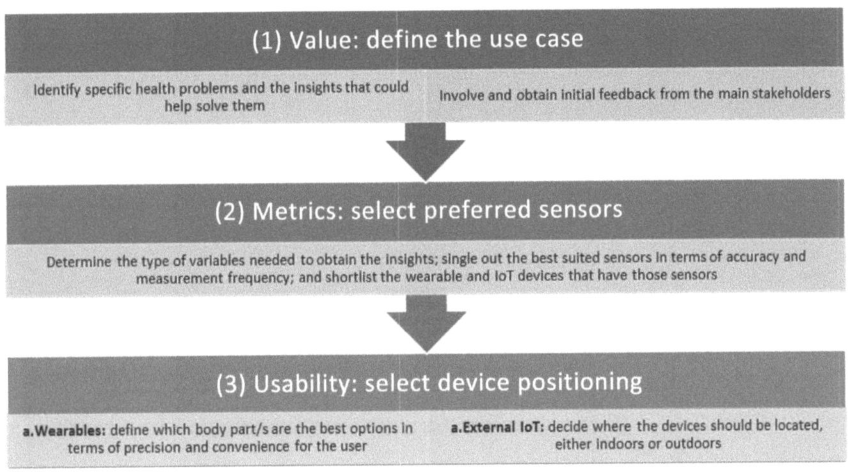

FIGURE 6.3

Main steps to implement wearable devices.

However, as opposed to other solutions and technologies, wearables require a greater degree of engagement from the stakeholders involved in its implementation. The main reason is that there is an almost limitless number of analytics that could be obtained or inferred by combining wearable data with data from other IoT devices or external sources such as, in this case, meteorological sensors. Therefore the feedback from key stakeholders, including final users; wearable hardware, software and analytics providers; health and healthcare professionals; and other decision makers, should be considered from the beginning. Multidisciplinary teams should define the general roadmap to collect data with wearable devices. Also, it is important to take into account from the beginning the target insights and added value.

Once the context and the added value of the wearable-driven use case are clear, the next step is to determine which physiological and atmospheric variables should be measured, and the best way to obtain reliable signals in real time. Nowadays, there are many wearable devices available in the market for various applications. Each of them has a set of features, but their general capabilities could be classified according to the type of signals they provide. Fig. 6.4 shows an overview of some of the most common physiological and ambiance-related signals that can be obtained in real time with wearable sensors [18].

Signals from wearable sensors include well know parameters such as heart rate, respiration rate, body temperature (skin or core) and blood pressure; and more advanced signs such as electrocardiogram, a graph of voltage versus time that measures electrical activity of the heart; oxygen saturation, an estimate of the amount of oxygen carried by blood cells; and skin perspiration, a measurement of the electrical conductance of the skin used to infer the quantity of sweat produced by sweat glands. In addition to physiological signals, wearable sensors can also be used to evaluate a broad array of motion and ambiance parameters such as activity cycles or movement types; and temperature, light, humidity, noise, atmospheric pressure, and certain gases and chemicals respectively.

In most cases, the accuracy of the signals captured by wearable devices does not only depend on the method applied to obtain it (i.e., heart rate from the R-peak of the electrocardiogram vs photoplethysmography). The location of the wearables and their sensors is very important from a usability point of view as

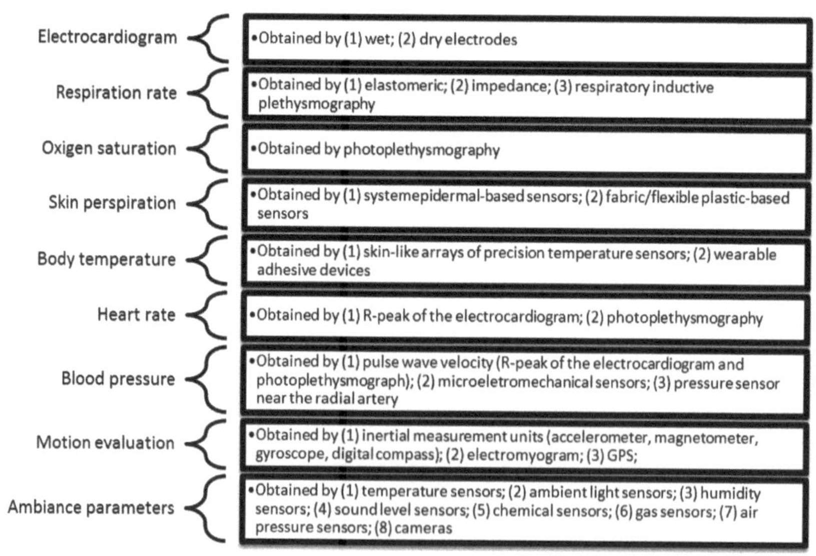

FIGURE 6.4

Overview of physiological and ambiance-related signals from wearable sensors.

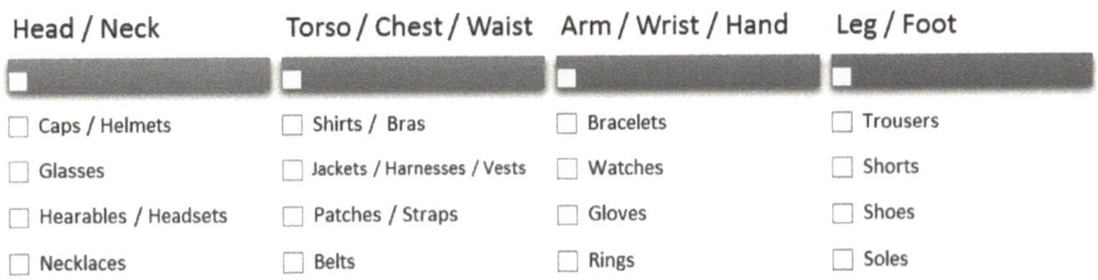

FIGURE 6.5

Classification of wearables devices by body part.

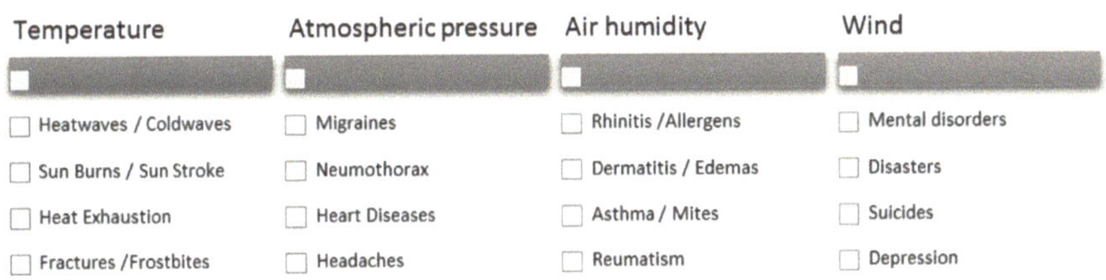

FIGURE 6.6

Connections between meteorological variables and diseases.

well as for data collection. Fig. 6.5 presents a classification by body part of the most common smart clothes or accessories that include one or multiple sensors [19] and can capture and transmit data in real time.

In addition to wearables, IoT in the environment can be very useful to provide contextual information on how external variables such as heat, air humidity, atmospheric pressure, solar radiation, atmospheric electricity, wind speed, air pollution, and allergens levels can alter, individually or in a synergetic way, citizens' physiological and psychological wellbeing.

6.2.2 Biometeorology and health

Biometeorology has studied the interactions between atmospheric processes and living organisms for decades. The knowledge obtained through this research can be used to define the limits where atmospheric conditions become a hazard for human health. Fig. 6.6 presents a brief example of the links between meteorological factors and different groups of diseases. As it can be seen, atmospheric factors and diseases are generally connected in a complex way. Despite the complexity, the combined use of data from wearable devices and biometeorological forecasting can be an excellent way to provide health services to high-risk demographic groups such as the elderly or those who suffer from weather-related chronic diseases [20].

Biometeorological studies analyze the interaction between the environmental conditions surrounding a person and his/her thermoregulation, acclimatization, and adaptation processes [21]. Thermoregulation is a continuous process that can vary from second to second whereas acclimatization takes much longer, from several weeks to months. For instance, acclimatization is triggered when a person travels to a country with a different climate or to a place located at a very different altitude. Biological adaptation refers to a transformation that affects genetic factors and is developed over a much longer period, up to a million years.

From a biometeorological point of view, the concept of health is a mix of an ecological and a medical approach, because health and wellbeing depend on the surrounding physical and social ecosystems, out of which thermal environment is a key component [22]. Fig. 6.7 presents how a meteorological variable and a human interact with each other during thermoregulation. Zone symbolizes the comfort area, which is different from person to person, where individual wellbeing is high about a specific meteorological factor (i.e., sun radiation). Area D1 shows the values where biometeorological distress increases due to lower exposure to the meteorological factor (i.e., sun radiation). The D2 area symbolizes a different type of distress due to greater exposure to the sun. Wearable devices can help to define the specific values where risk increases, either in one direction or another, because, for instance, too much solar radiation can produce skin burns and too little can cause rickets.

In practice, the combined use of wearables, medical devices, and meteorological sensors for biometeorological studies generates big and heterogeneous databases that contain medical, physiological and meteorological data. These data can be stored in the biometeorological data infrastructures developed by Fdez-Arroyabe [23]. These heterogeneous databases are used to study how thermal environments impact human health and wellbeing. For example, the Geobiomet Research Group created a mobile application (Oxyalert Beta) [23] that enables

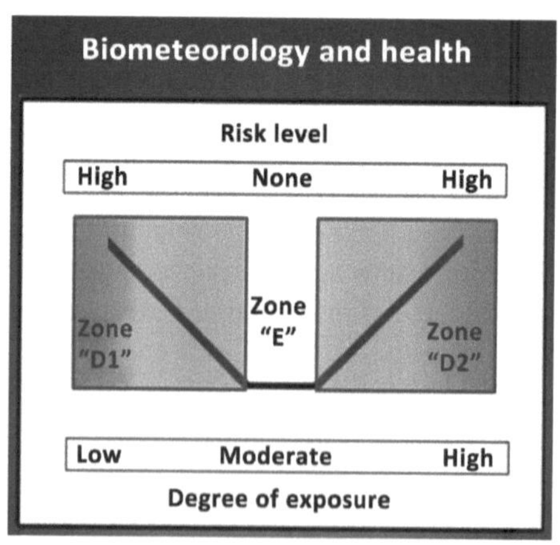

FIGURE 6.7

Physiological response to meteorological factors exposure.

users to test how meteor-sensitive they are to atmospheric oxygen changes. In addition, the app provides individualized 3-day forecasting of hypoxia and hyperoxia conditions for their geographic location. In this particular case, physiological changes of user have not yet been monitored through wearables but by self-answered questionnaires.

6.3 Heat stress and physiological mechanisms

6.3.1 The homeostatic process: a medical approach

Before explaining the physiological mechanisms and the systems that control body temperature, it is important to explain homeostasis. According to the Aristotelian conception of virtue, a virtuous man is one who is capable of achieving the equilibrium in his life. That could also apply to the human body. However, instead of equilibrium, physiologists named it homeostasis: the balance or the intent of the body to achieve "physiological virtue" [24]. The most common example of homeostasis related to temperature is gain or loss of body heat. The inability of the body to reach a homeostatic state can cause health problems.

For temperature homeostasis, the main feedstock is heat [25]. The concept of homeostasis applies to every physiological event of the human body related to a physical factor, and thus to multiple daily scenarios such as the following of Heyleen and Josh.

Heyleen is working on the streets of Ottawa. It is a very rainy day. She feels cold even though she is not wet and is wearing appropriate clothing. She feels a very cold sensation in her hands like there is no blood flowing through them; in fact, they look a little bluish. She continues to shiver until she goes to a café to meet some friends.

Josh is a middle-aged man working on a sugarcane plantation in Brazil. He has been cutting down cane plants under the sun for the last 2 hours. Suddenly, he realizes that his t-shirt is wet, especially the back. He decides to sit down next to some coworkers under a big umbrella and drink a bit of water. He looks at his hands and shows them to his fellow workers because the vessels are "too big."

In both situations, the self-regulating temperature mechanisms of the body were in action. In fact, there were multiple physiological responses to adapt the core temperature [26,27] of the body to the environment (Fig. 6.8).

- Cold environment

  Heyleen's hands were bluish because she experienced peripheral vasoconstriction, meaning that less blood was flowing to her hands. The diameter of her arteries decreased to keep all the heat in the "core body." In addition, she was shivering because she experienced a clonal contraction of her muscles to produce heat.

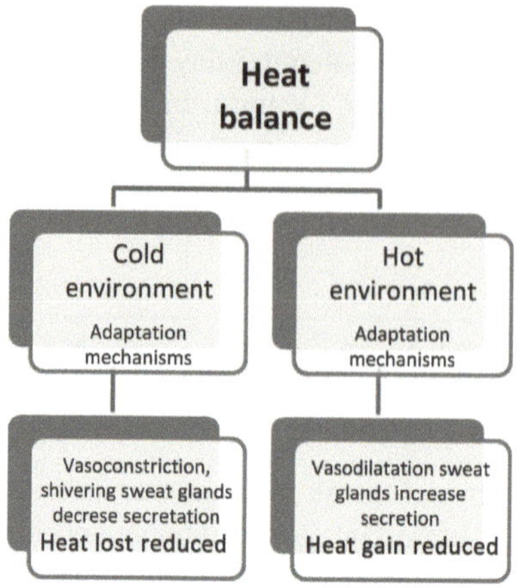

FIGURE 6.8

Thermoregulation process related to heat.

- Warm environment

Josh was sweating because his body was trying to reduce heat through the sweating system. The "too big" vessels on his hands were due to vasodilatation. Compared to Heyleen's vasoconstriction, vasodilatation increases the diameter of the vessels to reduce body heat.

The examples of Heyleen and Josh show that there is a continuous regulation of body temperature depending on the atmospheric conditions. Therefore the physiological mechanism of thermoregulation works to keep us alive in our ever-changing closest environment.

6.3.2 Heat control systems and pathways

These physiological mechanisms of thermoregulation involve multiple systems and body parts [24,25]. The hypothalamus and central nervous system (CNS) form the "head" of the thermoregulation system. The hypothalamus is divided into the anterior and the posterior parts.

The anterior hypothalamic-preoptic area has direct control over the sweating and vasodilatation mechanisms of the entire body. Consequently, the anterior hypothalamus can reduce body heat, and is triggered under high atmospheric temperature.

The posterior hypothalamic area is centered on the "integration." It receives the information transmitted to the anterior hypothalamic-preoptic area and other body parts. If this hypothalamic area does not function properly, which is relatively common in the elderly, the cells of the hypothalamus do not know if the atmospheric temperature is too high or too low. This can have serious implication on elderly morbidity and mortality when there are extreme conditions such as the ones during the heat or cold waves in Europe [29].

The second component is the peripheral nervous system (PNS). Its main function is to be the "informer" of the hypothalamus. PNS is related to the transmission of information between the CNS and the skin. It can be divided into two main functions.

First, environmental data compiling about the environment to inform the hypothalamus. This task is done by the afferent nervous cells, which integrate all the information given by the skin receptors into the central nervous system. Second, activate the sweating and shivering "processes." By the efferent neurons, the PNS can transmit the orders given by the hypothalamus to the muscles or skin.

The skin is part of the integumentary system. The skin plays a critical role in the process of heat exchange. There are two important parts of the thermoregulation system under the skin. On the one hand, surface receptors interpret the atmospheric environmental temperature and codify that information into electric and chemical signals. Then, the "codified" information can be transmitted through the nervous system. On the other hand, surface sweat glands are integrated into the skin in a way that when the electric signal arrives from the hypothalamus, usually from the anterior hypothalamic-preoptic area, the glands are activated to reduce body heat. Only when that information arrives into the hypothalamus, can we perceive a certain sensation [30,31]. This sensation is the sum of all the electrical signals triggered by the nerves and its receptors, and thus it lacks "consciousness" or "perception." Perception is understood as a subjective reading of a sensation [32]. Everyone experiences his or her own subjective reading of atmospheric temperature at any time what has allowed us to define concepts such as perceived or apparent temperature.

Environmental adaptation starts with changes in our receptors. A continuous input on a specific receptor can deactivate it, and the new input can be interpreted as normal [33]. For example, being exposed to a high-temperature environment for a long period can lead to the acclimatization and adaptation of that organism with that specific environment or thermal comfort [25]. There are two types of thermoregulation processes: the autonomic and the behavioral [28,30]. The autonomic process involves all the autonomous physiological aspects mentioned above, and the behavioral can be explained by the fact that if someone is cold, he or she will seek a warmer place [33,34], which depends on an individual's perception of thermal comfort.

6.4 Wearables, heat stress, and the workplace

6.4.1 Heat at work: a global concern

The International Labor Organization (ILO) is concerned about the impact of extreme heat on outdoor workers in low and middle income countries [35]. This is increasingly important because due to climate change, heat is expected to become a hazard for human beings. Thermal environment is a very relevant factor because there are multiple social and economic implications of the impact of heat and cold on living organisms. Thus thermal comfort and heat stress are two concepts that should be considered by authorities in relation to occupational safety and health [36] in different economic sectors such as agriculture, fishing, construction, mining, manufacturing, and urban areas.

Indoors environment is also important for the WHO and the ILO because in many cases, indoor locations are even five times more polluted than outdoors areas [36,37]. Artificial heating and cooling are also poorly regulated, and the consequences are unsustainable usage of energy, potential loss of productivity at work, and the spread of respiratory and vector-borne diseases such as colds, influenza, and allergies.

To unleash the potential of wearable devices and their adoption, it is important to define the use cases where they can add the most value. Health promotion and disease prevention at the workplace could be a very good starting point because, in the EU alone, every year more than 4000 workers die due to accidents and illnesses at work and more than 3 million suffer from at least 3 days of absenteeism due to occupational safety and health problems. Also, recent surveys show that 24.2% of workers consider that their health and safety is at risk due to their work, and 25% believe that work has a negative effect on their health [38]. On top of human suffering, the costs due to work-related sick leave are unacceptably high. For instance, in Germany, there is an average of 460 million days of sick leave per year, which leads to an estimated productivity loss of 3.1% of the GDP [39]. To solve this problem, it is important to integrate as many stakeholders as possible because each stakeholder has a unique point of view, different needs, and specialized knowledge that could help define the best wearable-driven solution possible. Fig. 6.9 shows the most relevant stakeholders that could benefit from and influence wearable use in the workplace.

Adding value to as many leads as possible in the chain of stakeholders is important because they are all highly interconnected and sometimes their interests could be in conflict. For instance, blue-collar workers could refuse to use wearable devices and share the data with their employers unless the benefits they obtain outweigh the potential loss of privacy incurred or even the hassle of using the device, whereas industrial companies might just be willing to implement wearable-driven solutions for worker health if they see a clear return on investment.

6.4.2 Wearables and heat indexes

Heat waves and the urban heat island effect are two relatively new concepts that explain why heat has become a source of environmental risk and is considered in terms of morbidity and mortality by nations and the WHO. Under this umbrella, there is global concern about how climate change and weather variability affect

FIGURE 6.9

Key stakeholders for wearable-driven services in the workplace.

human health. In addition to monitoring physiological parameters, which is essential, thermal comfort definition and heat stress assessment [40] is also a priority.

There are no traditional weather stations, nor satellite sensors, able to measure accurately and continuously the thermal environment at work, in the way the wearable and portable meteorological sensors do. Data collection of human physiology and air properties through wearables and IoT, at a very detailed spatiotemporal scale, represents a major opportunity to calculate thermal comfort indexes, define biometeorological risks, and outline human vulnerability to heat at work.

As has been shown, there is a wide range of wearables and medical devices measuring and monitoring physiological parameters. Fig. 6.10 shows three examples of wearables and portable devices that have been used to collect data in specific studies; from left to right: a basic wearable to monitor heart rate, sophisticated smart shirt with 1-lead ECG and motion sensors (accelerometer and gyroscope) developed by Ambiotex; and a couple of portable Velleman data loggers DVM171THD, with meteorological sensors, that measure relative air humidity and temperature at a very detailed temporal scale.

There are many indexes to assess how heat affects worker health, safety and productivity. A comprehensive catalogue and classification of human thermal climate indices with references to 162 indices were published by de Freitas and Grigorieva [41]. They vary depending on the number of variables considered, the rationale employed, and the relative sophistication of the schemes. A different

FIGURE 6.10

Wearable and portable devices to collect data.

study from Blazejeczyk et al. [42] correlated different thermal indexes with the Universal Thermal Comfort Index (UTCI) and concluded that those indices derived from human health heat budget models are strongly correlated with UTCI. In contrast, those based on relatively simple formulas showed weaker correlations. Lack of radiation factor in the equations was reported as one of the reasons to explain these weaker correlations. This information should be taken into account when designing new wearable devices in the future. Normally, temperature and air humidity sensors are included in portable devices but not sun radiation, which is an important factor for heat stress.

As mentioned by Kjellstrom et al. [35], one of the most commonly used indexes to protect workers from excessive heat at the workplace is the Wet Bulb Globe Temperature (WBGT) index. In this case, the body heat balance depends on six basic factors: air and radiant temperature, humidity, wind speed, clothing, and the metabolic heat generated by human activity. Due to climate change, heat at work creates an occupational health risk in low- and middle-income tropical countries [43].

According to Chuansi et al., the evaluation of occupational heat stress requires the measurement of four thermal climate factors [40] (air temperature, humidity, air velocity, and heat radiation) and the computation of the indexes can have a physics approach based on heat transfer and heat balance or an empirical one based on objective and subjective heat stress measurements.

The evaluation of heat stress must distinguish between occupational and general population heat stress, and the assessments can be individual or population based. It is also important to determine if the goal is just to assess heat stress in real time or also forecast future heat stress levels [40]. In any case, the difference between the calculated index value and the limit value must be considered. Some indexes of thermal discomfort are based on the creation of a catalogs of daily circulation weather types that are used to determine the relative risk of hospital admissions for specific diseases.

6.4.3 Case studies: outdoors versus indoors

6.4.3.1 Outdoors thermal comfort and heart rate

Outdoor thermal comfort varies depending on latitude and longitude. The geographic location where people usually work is very relevant to understand the physiological thermoregulation processes presented in previous section. Fig. 6.11 shows the temporal evolution of the heart rate of an outdoor worker, measured using a smart shirt with 1-lead ECG and motion sensors (accelerometer and gyroscope) from Ambiotex. Heart rate is expressed in beats per minute (bpm) at one-minute intervals.

Fig. 6.11 shows how heart rate values increase when workers face a demanding task in terms of physical activity. As can be seen in the graph, the worker reaches 140 bpm in some specific moments while in Fig. 6.12 a maximum rate reached was 110 bpm.

In this example, the heart rate varies. To identify risk patterns and develop a customized heat warning system, it would be necessary to elaborate a personalized study of the physical conditions of the worker, and this can only be done by recording his/her activity in different environmental conditions. In such a study, it would be important to distinguish how much of the distress could be attributed to workload and how much to thermal environment conditions.

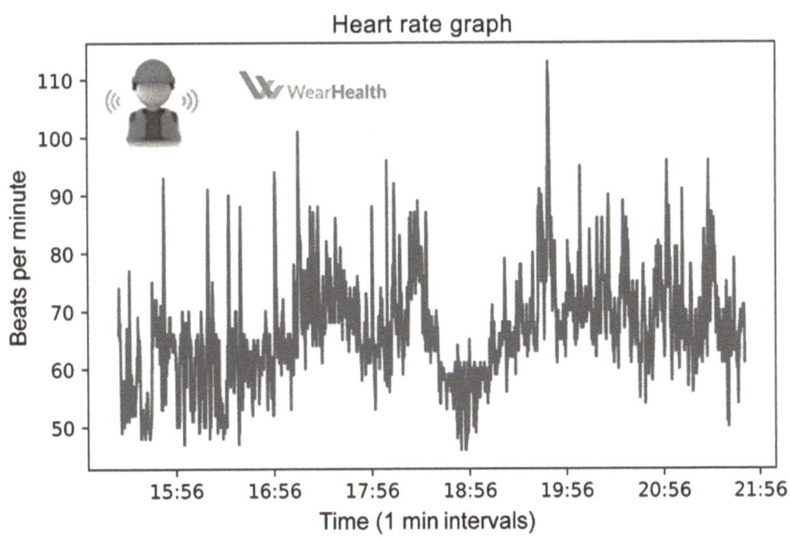

FIGURE 6.11

Heart rate of a worker during a basic activity.

WearHealth.

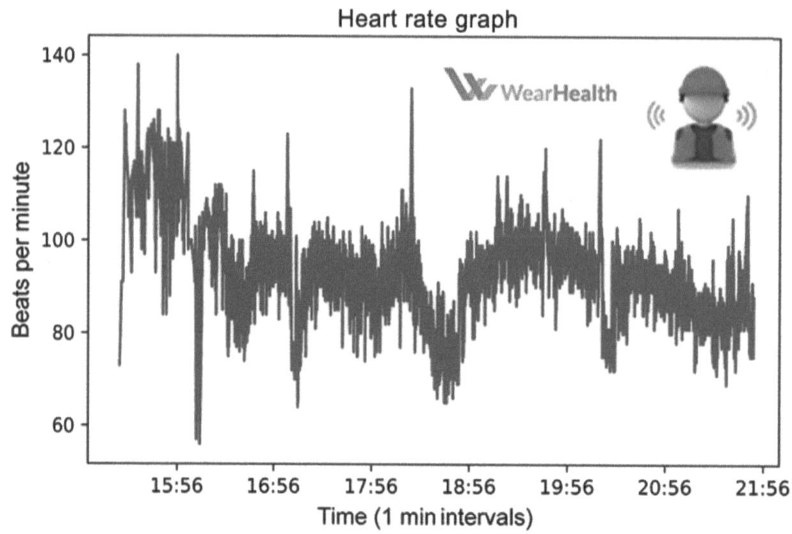

FIGURE 6.12

Heart rate of a worker during a physically demanding task.

WearHealth.

6.4.3.2 Indoors thermal comfort and heart rate

According to the Chartered Institute of Building Services Engineers of the University of Sonora the acceptable area of comfort for most types of work is between 16°C and 24°C. They recommend 16°C for factories, 18°C for hospitals, and 20°C for offices. However, in certain factories with high physical workload, 13°C could be a comfortable temperature, whereas very sedentary work could still be performed with reasonable comfort around 24°C [44]. The Health and Safety at Work Act specifies the conditions employers must provide employees for the development of working activities.

Two points are relevant in relation to heat stress at work. The first one is that employers must conduct risk assessments. The second one is that employers must pay attention to individual vulnerabilities. These two elements imply that there is a need for specific studies of biometeorological risks at work, and what is more important, a customized warning system at an individual level to define individual vulnerability to heat. The use of wearables plays a key role in defining occupational heat risks for workers.

Bioclimatic architecture and building designers control comfort at work inside buildings through the creation of psychometric charts that define the thermal comfort areas by combining dry and wet bulb temperatures with relative humidity. An example of a psychometric chart can be found in the study developed by Manzano-Agugliaro et al. [45] adapted from the Giovani [46]. Thermal comfort indoors is not only related to health and safety, it also affects sustainable development and employee productivity. In many cases, artificial heating and/or cooling inside buildings is not efficient in terms of energy consumption or worker productivity.

Puente and Fdez-Arroyabe [47] used the portable meteorological devices presented in Fig. 6.13 to collect data about the thermal conditions of the working environment of a group of professors. Data loggers collected information on temperature and air humidity in 17 rooms every ten minutes during several months corresponding with the autumn and winter seasons in Northern Spain.

The results showed that many offices reached the highest temperature on Friday afternoon at the end of the work week when the building was empty for the following 2 days. Moreover, there were big differences in temperature and air humidity from one room to another. The enormous variability registered in temperature and air humidity (Fig. 6.13) datasets in a small area of no more than 250 m^2 indicates how big spatial variability of thermal conditions can be indoors and the implications this can have in terms of use energy efficiently, productivity, and health and wellbeing at work. According to the records of the data loggers, there is one room that reached a minimum temperature of 11°C on January 27th and another room reached a maximum temperature of 39°C on February 31st.

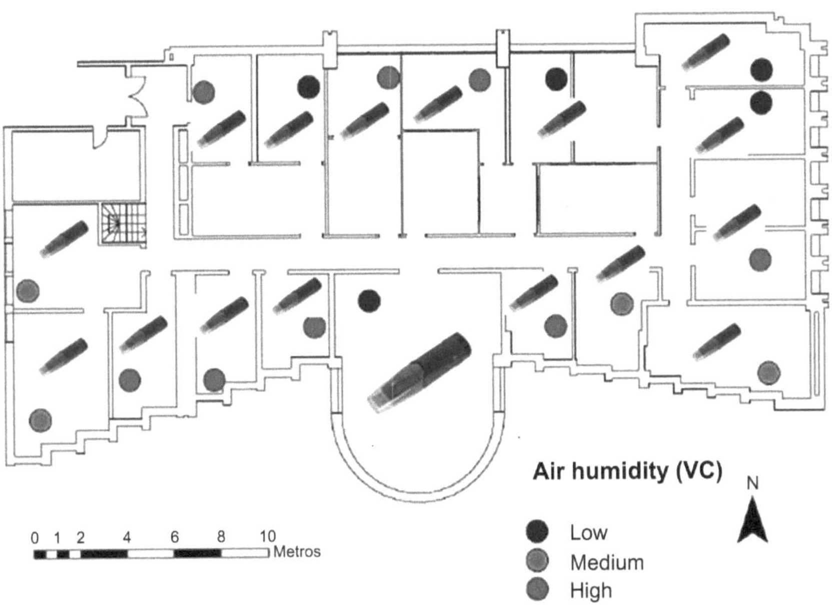

FIGURE 6.13

Variation coefficient of indoors air humidity.

From L. Puente, P. Fdez-Arroyabe, Indoors thermal comfort variability and its impacts on energy efficiency, health and productivity at work. UCREA-UC, 2016. Available from: https://repositorio.unican.es/xmlui/handle/10902/9285.

The range of air humidity in the study area went from 22.6% to 78.9% depending on the exposure and location of the office.

First, energy should be used more efficiently. Secondly, productivity can be affected by some specific thermal comfort levels registered in the study. It is more difficult to concentrate when it is very hot. In addition, some of these specific environmental conditions could promote the transmission of infectious diseases such as influenza or syncytial viruses among many others, mainly when it is cold. These three basic aspects show the importance of thermal comfort indoors.

6.5 Conclusion

The implementation of wearables requires a great degree of cooperation with the person involved in its implementation. It is a process of cocreation of science where all the stakeholders (employers, employees, institutions, public/private partnerships, governments, health services) must cooperate to reduce heat-related morbidity and mortality at work. The use of wearables must follow the exiting law for data collection in the country where the employees work and must be clear to avoid misunderstanding, especially those related to the fear employees may have in relation to the use of these data by employers to overcontrol their activities at work. Data protection guarantees must be also provided to users in relation to data privacy.

The concurrent registration of physiological parameters of individuals and thermal environment properties through wearables offers science the opportunity to study the atmospheric component of many weather-related diseases at a personal level. This can facilitate the development of climate-relate preventive medicine at work.

The massive use of wearables and medical devices will facilitate the formation of heterogeneous data bases at a personal level. In this frame, personalization of treatment based on a werable's real-time data approach can be offered to patients in order to facilitate a more efficient care to each individual.

In addition, advances in biophysical knowledge must be used to design new biosensors and include them in future wearables. According to the homeostatic processes mentioned here and the existing heat control systems of each person, customized biometeorological warning systems can be developed for workers at an individual level. With this implementation, the efficiency at work and workers wellbeing can be increased or, at least maintained.

References

[1] P. Kostkova, Grand challenges in digital health, Front. Public Health 3 (2015) 134.

[2] Soreon Research, The Wearable Health Revolution: How Smart Wearables Disrupt the Healthcare Sector, 2014.

[3] P. Kostkova, H. Brewer, S. de Lusignan, E. Fottrell, B. Goldacre, G. Hart, et al., Who owns the data? Open data for healthcare, Front Public Health 4 (2016) 7.

[4] X. Li, J. Dunn, D. Salins, G. Zhou, W. Zhou, S.M. Schüssler-Fiorenza, et al., Digital health: tracking physiomes and activity using wearable biosensors reveals useful health-related information, PLoS Biol. 15 (1) (2017) e2001402.

[5] Cisco Systems, Cisco Visual Networking Index: Global Mobile, 2016. Available from: <http://www.statista.com/statistics/487291/global>.

[6] K. Samydurai, Technology: a key to patient satisfaction? Manag. Healthc. Exerc. April (2016).

[7] D.V. Dimitrov, Medical Internet of things and big data in healthcare, Healthc. Inform. Res. 22 (3) (2016) 156–163.

[8] C.S. Bloss, N.E. Wineinger, M. Peters, D.L. Boeldt, L. Ariniello, J.Y. Kim et al., A prospective randomized trial examining health care utilization in individuals using multiple smartphone-enabled biosensors, Peer J. (2016) Available from: <https://peerj.com/articles/1554>.

[9] E. Silver, S. Wallenstein, A. Levy, Inward and outward: the role of patient self monitoring and patient communities: inflammatory bowel diseases, Inflamm. Bowel Dis 25 (2) (2012) 45.

[10] A. Kong, S.A.A. Beresford, C.M. Alfano, K.E. Foster-Schubert, M.L. Neuhouser, D. B. Johnson, et al., Self-monitoring and eating-related behaviors are associated with 12-month weight loss in postmenopausal overweight-to-obese women, J. Acad. Nutr. Diet. 112 (9) (2012) 1428–1435.

[11] M. Swan, The quantified self: fundamental disruption in big data science and biological discovery, Big Data 1 (2) (2013) 85–99.

[12] J. Norris, Self-tracking may become key element of personalized medicine | UC San Francisco, 2012. Available from: <http://www.ucsf.edu/news/2012/10/12913/self-tracking-may-become-key-element-\personalized-medicine>.

[13] S. Fox, M. Duggan, Mobile Health 2012, 2012.

[14] P. Fdez-Arroyabe, D. Roye, Chapter 9: "Co-creation and participatory design of big data infrastructures on the field of human health related climate services", Internet of Things and Big Data Technologies for Next Generation Healthcare, Springer, 2017, pp. 199–226.

[15] P. Fdez-Arroyabe, Climate services and human health: a niche of opportunities for economic growth, Sci. Ann. Alexandru Ioan Cuza Geogr. Ser. 59 (2) (2013) e2284–e6379.

[16] P. Fdez-Arroyabe, D.T. Robau, Past, present and future of the climate and human health commission, Int. J. Biometeorol 61 (1) (2017) 115–125.

[17] R.E. Herzlinger, Why innovation in health care is so hard. Harv. Bus. Rev., 2006. Available from: <https://hbr.org/2006/05/why-innovation-in-health-care-is-so-hard>.

[18] D. Dias, J. Paulo Silva Cunha, Wearable health devices—vital sign monitoring, systems and technologies, Sensors 18 (8) (2018) 2414.

[19] Vandrico Inc., Wearable technology database, 2018. Available from: <https://vandrico.com/wearables/>.

[20] P. Fdez-Arróyabe, Chapter 22: "Meteorological conditions and human health", Carlos Garcia-Legaz Martínez & Francisco Valero Rodríguez (Eds.) in: Adverse Weather in Spain, AVM Ediciones, Madrid, 2013, pp. 325–342.

[21] J. Smithers, B. Smit, Human adaptation toclimatic variability and change, Glob. Environ. Change 7 (2) (1997) 129–146.
[22] P. Fdez-Arroyabe, Climate change, local weather and customized early warning systems based on biometeorological indexes, J. Earth Sci. Eng. 5 (2015) 173–181.
[23] P. Fdez-Arroyabe, L. Lecha Estela, F. Schimt, Digital divide, biometeorological data infrastructures and human vulnerability definition, Int. J. Biometeorol. 62 (5) (2018) 733–740.
[24] J.E. Hall, Chapter 74: "Body temperature regulation and fever", Guyton and Hall: Textbook of Medical Physiology, thirteenth ed., Saunders, 2016, pp. 911–919.
[25] K. Nagashima, K. Tokizawa, S. Marui, Thermal confort, Handb. Clin. Neurol. 156 (2018) 249–260.
[26] S. Toy, N. Kántor, Evaluation of human thermal comfort ranges in urban climate of winter cities on the example of Erzurum city, Environ. Sci. Pollut. Res. Int. J. 24 (2) (2017) 1811–1820.
[27] E. Schulze, H.A. Daanen, K. Levels, J.R. Casadio, D.J. Plews, A.E. Kilding, et al., Effect of thermal state and thermal comfort on cycling performance in the heat, Int. J. Sports Physiol. Perform. 10 (5) (2015) 655–663.
[28] Y. Zhu, Q. Ouyang, B. Cao, X. Zhou, J. Yu, Dynamic thermal environment and thermal comfort., Indoor Air. 26 (1) (2016) 125–137.
[29] O. Lottka, J. Kyselý, Characterizing joint effects of spatial extent, temperature magnitude and duration of heat waves and cold spells over Central Europe, Int. J. Climatol. 35 (7) (2015) 1232–1244.
[30] G. Katavoutas, M.N. Assimakopoulos, D.N. Asimakopoulos, On the determination of the thermal comfort conditions of a metropolitan city underground railway, Sci. Total Environ. 566-567 (2016) 877–887.
[31] M. Rutty, D. Scott, Bioclimatic comfort and the thermal perceptions and preferences of beach tourists, Int. J. Biometeorol. 59 (1) (2015) 37–45.
[32] H. Oi, T. Hashimoto, T. Nozawa, A. Kanno, N. Kawata, K. Hirano, Neural correlates of ambient thermal sensation: an fMRI study., Sci Rep. 7 (1) (2017) 112–179.
[33] A. Mendes, A.L. Papoila, P. Carreiro-Martins, L. Aguiar, S. Bonassi, I. Caires, The influence of thermal comfort on the quality of life of nursing home residents, J. Toxicol. Environ. Health. 80 (13–15) (2017) 729–739.
[34] A. Kurz, Physiology of thermoregulation, Best Pract. Res. Clin. Anaesthesiol. 22 (4) (2008) 627–644.
[35] T. Kjellstrom, I. Holmer, B. Lemke, Workplace heat stress, health and productivity: an increasing challenge for low and middle income countries during climate change, Glob. Health Action 2 (2009) 46–51.
[36] VELOUX Group, Indoors generation video, WHO. First WHO Global Conference on Air Pollution and Health, 2018. Available from: <https://www.who.int/airpollution/events/conference/en/>.
[37] WHO, First WHO global conference on air pollution and health, 2018. Available from: <https://www.who.int/airpollution/events/conference/en/>.
[38] Eurostat, Accidents at work statistics – Statistics explained, 2018. Available from: <https://ec.europa.eu/eurostat/statistics-explained/index.php/Accidents_at_work_statistics>.
[39] Eurofound, Absence from work – Germany, 2010. Available from: <https://www.eurofound.europa.eu/publications/report/2010/absence-from-work-germany>.

[40] G. Chuansi, K. Kalev, Ö. Per-Olof, K. Tord, Occupational heat stress assessment and protective strategies in the context of climate change Special Issue on Transdisciplinary Approaches to Climate Change, Int. J. Biometeorol. 62 (3) (2018) 359–371.

[41] C.R. de Freitas, E.A. Grigorieva, A comprehensive catalogue and classification of human thermal climate indices, Int. J. Biometeorol. 59 (1) (2015) 109–120.

[42] K. Blazejczyk, Y. Epstein, G. Jendritzky, H. Staiger, B. Tinz, Comparison of UTCI to selected thermal indices, Int. J. Biometeorol. 56 (3) (2012) 515–535.

[43] T. Kjellstrom, B. Lemke, M. Otto, Mapping occupational heat exposure and effects in South-East Asia: ongoing time trends 1980-2009 and future estimates to 2050, Ind. Health 51 (2012) 56–67.

[44] UNISON, Healthy and Safety Information Sheet, Temperature at work, 2014.

[45] F. Manzano-Agugliaro, F.G. Montoya, A. Sabio-Ortega, A. García-Cruz, Review of bioclimatic architecture strategies for achieving thermal comfort, Ren Sust En Rev. 49 (2015) 736–775.

[46] B. Givoni, Comfort, climate analysis and building design guidelines, Energ Build. 18 (1) (1992) 11–23.

[47] L. Puente, P. Fdez-Arroyabe, Indoors thermal comfort variability and its impacts on energy efficiency, health and productivity at work. UCREA-UC, 2016. Available from: <https://repositorio.unican.es/xmlui/handle/10902/9285>.

Further reading

T. Kjellstrom, Impact of climate conditions on occupational health and related economic losses—a new feature of global and urban health in the context of climate change, Asia Pacific J. Pub. Health 28 (2) (2016) 28–37.

CHAPTER

Reading Assistant: a reciter in your pocket

7

Chintan Bhatt, Pooja Kothari, Shweta Masrani and Vidhi Kokel
Charotar University of Science and Technology, Anand, India

Chapter Outline

7.1 Introduction

"For people without disabilities, technology makes things easier. For people with disabilities, technology makes things possible."

IBM training manual 1991.

Technology is changing the world drastically. In *The Nature of Technology: What It Is and How It Evolves* William Brian Arthur says that new technologies are put together from existing technologies. Thus, collectively technology evolves by creating new elements from within itself and can be seen as an evolutionary pattern. This evolution has created a far superior, more agreeable, and simpler life for people.

Reading Assistant is a wearable device that helps people read books and other reading material that might otherwise be difficult or impossible to read. While there is reading software available to access digital information, nondigital and nonbraille text are inaccessible.

Other such reading devices have been developed. For example, orcam, which is a device that can be attached to spectacles of the user and when the user points

Wearable and Implantable Medical Devices. DOI: https://doi.org/10.1016/B978-0-12-815369-7.00007-0

to some text using his/her finger, it reads the text aloud. Various mobile applications are also available that can capture text using the camera and then read it aloud.

The Reading Assistant discussed in this chapter, captures the image, extracts the text from it, and reads it aloud. Since it provides an audio file of the printed text, it can help visually impaired people or people with some kind of visual disorder read physical books.

Internet of Things (IoT) can be essentially defined as a network of physical appliances embedded with electronics, sensors, etc., that communicate with each other that is, collect and transfer data to each other to solve some specific purpose. Devices connected to one another through IOT can be detected or supervised remotely resulting in improved productivity, precision, and economic benefits in addition to requiring less human interaction [1].

Following is the analogy of the Reading Assistant with layers of an IoT architecture.

1. Perception layer: This layer usually comprises sensors, camera, monitoring stations, etc., to collect information. Reading Assistant consists of the camera which captures an image and passes it to the raspberry pi for processing through the wired connection between the camera and pi.
2. Network layer: This layer is from where the information is passed through different wired or wireless networks like the Internet. Hence in Reading Assistant the connection of the Raspberry pi with the Internet for processing of images is considered to be a part of this layer.
3. Application layer: This layer consists of Reading Assistant's module that processes the image, extracts characters from it, and reads them.

The World Health Organization estimates that about 285 million people are visually impaired and 39 million of them are blind. Furthermore, an estimated 246 million people live with moderate to severe vision impairment. According to one study 80% of what we learn depends on what we observe and the competence of our visual system [2].

Reading requires efficient and prompt communication between areas of the cortex that handle visual, hearing, and language information. If the sensory system for vision declines to quickly and precisely recognize tiny imprinted symbols (i.e., letters) then any endeavor to preach reading skills with traditional perspective will have little achievement.

However, according to a study, when a sense is missing, there is a fair amount of compensation by the other senses. There are some scientific research and facts which shows how hearing can be useful for visually impaired to learn effectively. Visually challenged individuals have a solid hearing sense that encourages them to acquire information and sort their thoughts orally [3].

Bradley-Johnson suggested, "Visually impaired students are not more sensitive than sighted students in terms of their ability to discriminate using hearing, taste, touch, and smell. Instead, they attend better when receiving information via these senses and are thus better able to interpret this information" [3].

Dyslexia is a disorder that causes a person difficulty of connecting the letters they see on a page with the sounds those letters and combinations of those letters make. Thus, people having Dyslexia struggle to read with fluency. Dyslexia affects 20% of the population and affects 80%–90% of people with learning disabilities.

Thus, this device can prove to be of great significance for blind people as Braille language version cannot be found for every book everywhere, for people with weak vision and for people suffering from dyslexia disorder. Apart from this, it can even act as a device for interactive learning at educational institutes and many more such applications can be drawn from it.

7.2 Related work

Available literature on image-to-speech research is summarized as follows.

Orcam is a tiny device developed by a startup in Israel that fits on a pair of glasses and reads printed text from pamphlets, business cards, products, etc. It even helps users recognize faces (previously stored in a database), money notes, and colors of clothes. It also gives the current time and date. This device is invoked by a finger pointing at something and then Orcam does its work. But as this device becomes active using finger gesture the device might start reading undesirable things at undesirable times.

The research papers "FingerReader: A Wearable Device to Support Text Reading on the Go" and "FingerReader: A Wearable Device to Explore Printed Text on the Go" describe a device that can be worn on an index finger and uses Tesseract OCR (optical character recognition) and Flite TTS(text-to-speech) to turn printed text into audio. The FingerReader hardware elaborates on the EyeRing device by appending multimodal feedback using vibration of motors. These vibration motors signal when the user's finger has reached the end of a line. This device does a local-sequential scan, providing continuous feedback and nonlinear text skimming and then takes a close-up image of printed text under the index finger or that is pointed to by it, processes it into its software stack and reads the printed text aloud. This device includes a desktop application for Windows and Mac operating systems. The response time of this device has been optimized to some extent and can be considered as a prominent feature of the device. It is user friendly but since it reads continuously word by word, no audio file is generated or stored for further use [4–6].

The research paper "Camera Reading for Blind People" describes an IOS application that uses OCR and TTS technologies to help blind people read printed materials. The architecture of this project consists of stages like image capture using a mobile device (with the iOS operating system), preprocessing of the image, OCR framework for recognition of text from the image, postprocessing and synthesis of the acquired output using the TTS module. Here, the

Levenshtein distance algorithm was used to compare the recognized text with the original text. A major shortcoming of this product is that it is IOS platform-specific and cannot be used on other platforms. While the preprocessing and post-processing of captured images has been notably beneficial for better results. [7]

The Phone Reader is an Android application. It allows the user to take a picture using the camera of an android phone, uploads it to a server where it is processed using Imagemagik commands. This image is then passed to the Tesseract OCR module that performs OCR, extracting text from the image and generates a text file. The extracted text file then goes into an Apertium translator module that translates the text file into the language chosen by the user (this step is optional). After the text has been translated on the server, the text file is automatically downloaded to the device and is processed using the Android TTS engine to create the audio for the text file. Phone Reader even allows the user to crop the image and read only a specific portion of a page. It also stores the extracted text file on the mobile device for later use, but it does not store the audio file. Also, the OCR module of this device is accurate only up to 80% that too for a quality image [8].

Although there are numerous studies and applications that address the theme of this work, the available work has limitations like low efficiency in character recognition, especially in conditions where image capturing quality does not match the desired level. This research area thus requires achieving prominent development to surpass the shortcomings of reality where there are improper light conditions, hazy focus, misalignment or the presence of shadows, which leads to ineffective recognition by OCR.

With Reading Assistant, we have tried to overcome some of the mentioned limitations and others by creating a device that works on any platform, is wearable, and that stores the audio file for further use.

7.3 Reading Assistant architecture

Reading Assistant takes an image of printed text using the camera attached to the device, processes the image, recognizes the text in the image, and extracts the text thereby generating a text file that is used to produce audio that can be heard through the speaker. Fig. 7.1 shows the architecture of Reading Assistant.

As can be seen in the figure, the input image passes through different stages before an audio file is generated.

In order to make it a wearable device the following hardware component are needed:

- Raspberry Pi Model B: This is a miniature of computer with an efficient processing speed. Due to its small size it is handy in nature.
- Pi Camera version 2: This is the most compatible camera with the Raspberry Pi. It is small in size and is capable of recording high-definition videos and

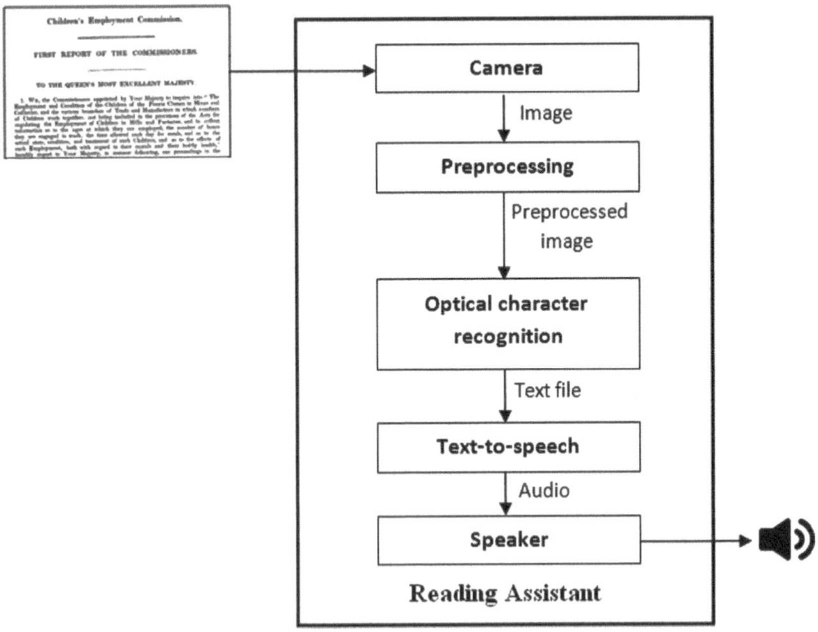

FIGURE 7.1

Architecture of Reading Assistant.

images. It can also be used for time-lapse and slow-motion video. It is compatible with all models of Raspberry Pi and has the following specifications:
- Small board size: 25 mm × 20 mm × 9 mm.
- A 5 MP (2592 × 1944 pixels) Omnivision 5647 sensor in a fixed focus module.
- Pi Camera support 1080p30, 720p60 and 640 × 480p60/90 resolution video record.

- Compatible Speakers:
  - Bluetooth speaker: Any speaker with Bluetooth connectivity can be configured with the device. This makes Reading Assistant compatible with a wide range of Bluetooth speakers.
  - Headphones or earphones with 3.5 mm audio jack

As seen in the architecture of the device; the preprocessing module, OCR module, and Text-to-Speech module are implemented in a single unit that is deployed onto a Raspberry Pi with which the camera and speakers are integrated or connected.

The device works as follows. First, it takes an image of the printed text with the camera that is integrated with the Raspberry Pi. The captured image is sent to the software module that is deployed on the Raspberry Pi. In the preprocessing module it is pre-processed and then passed to the OCR module wherein the text is extracted from the image and a text file is generated. This text file is then passed to the text-to-speech unit wherein it is converted to an audio file that is heard through the speakers of the device. The following sections describe the modules in more detail.

7.3.1 Preprocessing module

The image may or may not have a clear background, it could be de-skewed, or it might be hazy. To improve the image quality, the following preprocessing steps are taken [9]:

- Deskewing: A skewed image or an image that is rotated to some degree can be straightened using this function of the module [10].
- Conversion to greyscale: In this step, the image is converted to a greyscale format.
- Binarization: In this step, the input image is converted to a bi-level image and hence into a binary form. This process is known as binarization. The goal is to extract those pixels that represent objects such as text [11].
- Denoising: In this step, the noise of the image is reduced for efficient recognition.

These steps of image preprocessing are implemented using the OpenCV library, which is an open source library for image processing (Figs. 7.2–7.6).

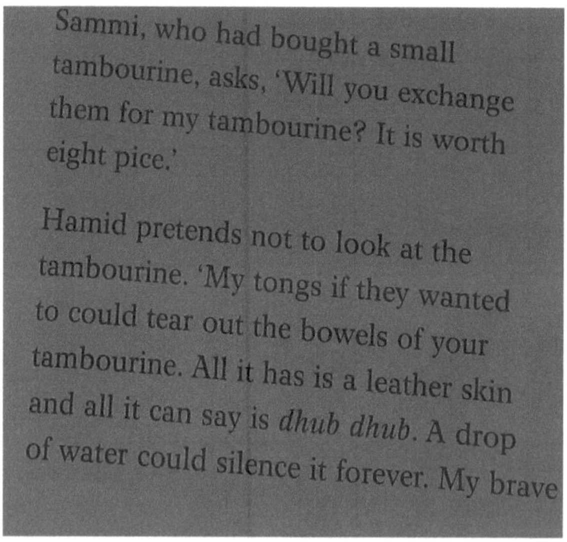

FIGURE 7.2

Original input.

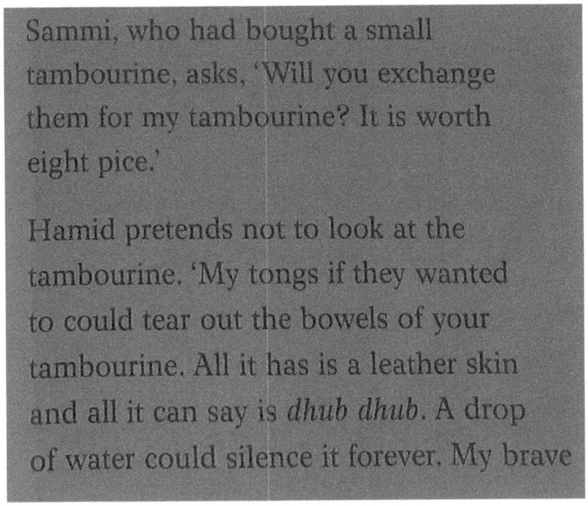

Sammi, who had bought a small tambourine, asks, 'Will you exchange them for my tambourine? It is worth eight pice.'

Hamid pretends not to look at the tambourine. 'My tongs if they wanted to could tear out the bowels of your tambourine. All it has is a leather skin and all it can say is *dhub dhub*. A drop of water could silence it forever. My brave

FIGURE 7.3

Deskewing.

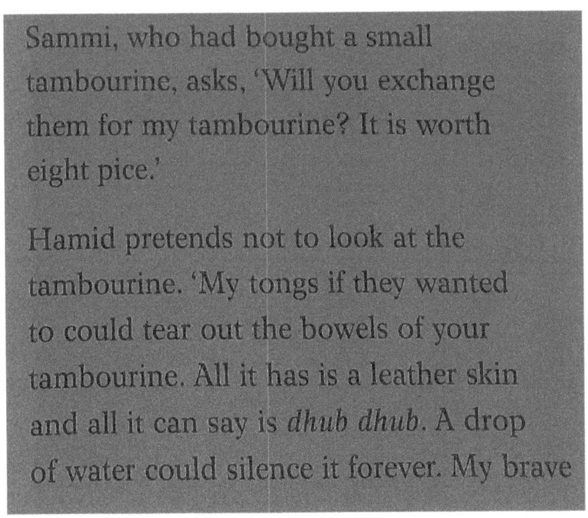

Sammi, who had bought a small tambourine, asks, 'Will you exchange them for my tambourine? It is worth eight pice.'

Hamid pretends not to look at the tambourine. 'My tongs if they wanted to could tear out the bowels of your tambourine. All it has is a leather skin and all it can say is *dhub dhub*. A drop of water could silence it forever. My brave

FIGURE 7.4

Grayscale image.

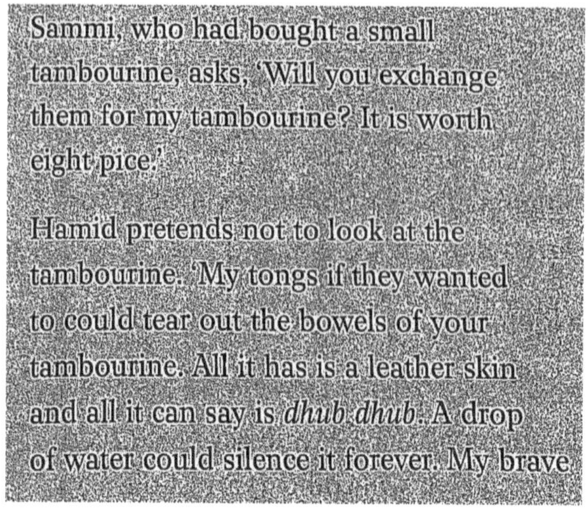

Sammi, who had bought a small tambourine, asks, 'Will you exchange them for my tambourine? It is worth eight pice.'

Hamid pretends not to look at the tambourine. 'My tongs if they wanted to could tear out the bowels of your tambourine. All it has is a leather skin and all it can say is *dhub dhub*. A drop of water could silence it forever. My brave

FIGURE 7.5

Binarization.

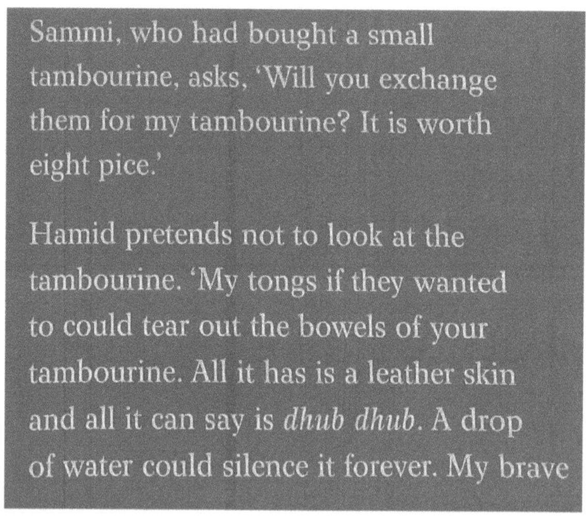

Sammi, who had bought a small tambourine, asks, 'Will you exchange them for my tambourine? It is worth eight pice.'

Hamid pretends not to look at the tambourine. 'My tongs if they wanted to could tear out the bowels of your tambourine. All it has is a leather skin and all it can say is *dhub dhub*. A drop of water could silence it forever. My brave

FIGURE 7.6

Denoising.

7.3.2 Optical character recognition

Optical character recognition OCR, just as its name implies, recognizes text from images and converts it to a text document. It is a process in which text is "extracted" from images or PDFs. This technology has helped minimize the need of manual effort to retype the text embedded in images [12].

Our device uses tesseract, an open-source OCR library by Google, to perform OCR on the input image. To improve accuracy, the images are first pre-processed and then passed to the tesseract module.

The main concept behind recognition of structural arrangement of alphabets is to train the device with their different patterns in which they occur and with their appearance. In OCR such orders are alphabets, numbers and some special characters like commas, question marks, etc. The device is trained by feeding various specimen of characters belonging to distinct sets or categories. According to this training, the software generates a model and forms specifications of every single character. In the second phase of recognition, untrained models of a dataset are fed into the machine. Then, the machine compares this data with the trained data and finds the best match for that particular character (Fig. 7.7).

7.3.2.1 Image scanning

In this process, an image is captured through a camera or scanner and provided as an input to the device. After that it is sent to preprocessing module of our device.

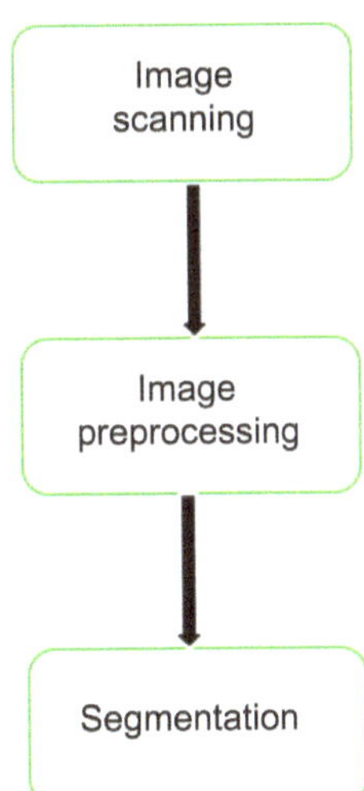

FIGURE 7.7

Components of optical character recognition.

```
Mggatmnlljswho had bought a small
tambourine, asks, `Will you exchange
them for my tambourine? It is worth
eight pice.'
Hamid pretends not to look at the
tambourine. `My tongs if they wanted
to could tear out the bowels of your
tambourine. All it has is a leather skin
and all it can say is dhub dhub. A drop
of water could silence it forever. My brave
```

FIGURE 7.8

Output.

7.3.2.2 Segmentation

In this process the printed characters are identified within the document and differentiated from figures and graphics [10]. Segmentation is performed in three steps:

1. Line segmentation
2. Word segmentation
3. Character Segmentation
   - Line Segmentation: This is the main step of the segmentation process. The lines of text are separated using horizontal projection. Image from which the text is to be extracted, is scanned from pixel-row to pixel-row. Histogram is plotted of these pixels. Through this the pixels are grouped according to their values and the lines are segmented from the image.
   - Word Segmentation: Next, word segmentation takes place. In this process individual words are extracted from lines. The gaps between the words are calculated from the vertical projection profile.
   - Character Segmentation: Segmentation of characters becomes easy when each character has the same breadth. The character segmentation gets more difficult when the breadth of the symbols depends on their structure ("proportional pitch"), when spacing, cursive or italic characters are present or due to less space the characters are touching each other (Fig. 7.8).

7.3.3 Text-to-speech module

The last is the text-to-speech synthesis. It is a method that converts text input to voice output. Google text-to-speech (gTTs) is a text-to-speech library that converts text documents to voice. Reading Assistant uses this library. gTTs allows unlimited lengths of text to be read, maintaining proper intonation and using abbreviations. Also, the sound of gTTs is similar to the natural voice of a human beings so people can understand it more easily. gTTs first parses the text and then saves it as an .mp3 file for later use. Thus, if a user wants to read a particular text

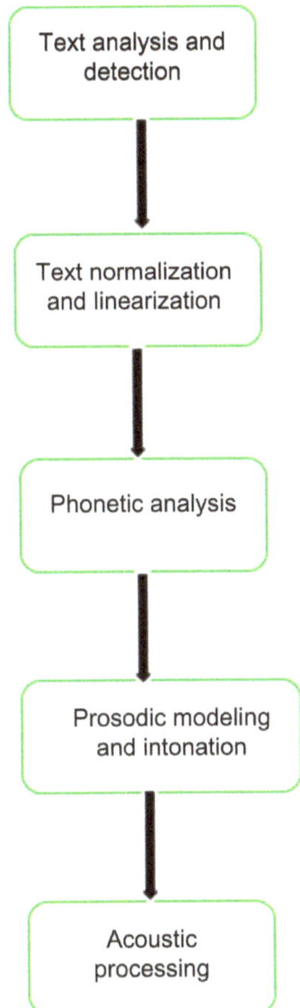

FIGURE 7.9

Components of text-to-speech.

file again, user has to just play that mp3 file instead of following the whole process. Components of the text-to-speech system are [13]:

- Text analysis and detection
- Text normalization and linearization
- Phonetic analysis
- Prosodic modeling and intonation
- Acoustic processing (Fig. 7.9).

7.3.3.1 Text analysis and detection

In text analysis, the text is parsed and divided into a list of words known as tokens. POS tagging means correctly identifying whether the extracted word is either a noun, verb, or either of any seven parts of speech. Part of speech is a process of assigning correct POS tags to each word in a sentence from a given set of tags that helps in understanding the real context of a sentence.

7.3.3.2 Text normalization and linearization

It is the conversion of text into pronounceable form. The purpose of this process is to detect punctuations and pauses in text. Generally, it is done to convert uppercase to lowercase, remove punctuations and to interpret correctly the same sounding words. It also differentiates between how to pronounce a number in different situations such as if it is a date or if it is a phone number. Identifying the correct pronunciation in each context is done through this method. Known abbreviations are converted to full form like Mr to master.

7.3.3.3 Phonetic analysis

According to the study of phonetics, a phone is any well-defined sound, irrespective of whether the literal sound is pivotal to the connotation of words. In this process, conversion of structural symbols into their respective phonetic sounds takes place. In TTS (text-to-speech) a word is pronounced in two ways:

- Dictionary way
- Rule-based way

The dictionary way is faster since the pronunciations and spellings are already stored. This approach can be used for words already present in the dictionary.

If we encounter unknown words in the text, then we use a rule-based approach to determine their pronunciations applying some rules on them that are defined in the system to find out their correct sound. This method does not need extra space for storage of words like in dictionary method.

7.3.3.4 Prosodic modeling and intonation

Prosody refers to the combination of stress, pattern, rhythm, and intonation in speech. This module depicts the emotions to be used. Intonation refers to pitch, which is used to express emotions such as happiness, questions, and exclamations. Different intonations such as rising, falling, dipping, and peaking are used to bring out different emotions in speech. All dialects use pitch, as intonation to project an example, to emote joy, or for questioning. This module helps in increasing the quality of speech and to bring naturalness to it.

7.3.3.5 Acousting processing

Here the voice is synthesized according to the speech characteristics of a person. Different types of synthesizing methods include:

1. Concatenative synthesis: This method is based on the joining or linking of fragments of recorded audio in human voice. It maintains a database of pre-recorded words [13]. Although, it is like a natural-sounding voice, it has an overhead of database due to which this synthesis is less efficient.

2. Articulatory synthesis: This method uses the physical prototypes of the system that is similar to how humans perceive speech. It uses physical and aural models of speech production to process speech. This technique uses five articulatory specifications: opening of lip, tongue blade, glottal area, nasal cavities, and vocal tract.
3. Formant synthesis: Also defined as rule-based synthesis, this method generates speech using rules that correlate to various sounds. The output speech here is clear, not only at normal speed but when the velocity is high. The only disadvantage here it is that the sound is robotic instead of natural.
4. Domain-specific synthesis: Just like its name implies, the outcome of synthesis is specific to that particular domain itself. It integrates recorded words and phrases to generate comprehensive statements [14]. This method is commonly used in the field of weather prediction, for schedules of transit systems, in talking watches and calculators, etc.

7.3.4 Raspberry Pi

All the modules explained above are integrated into a single program that is deployed onto a Raspberry Pi. This minicomputer has its own operating system,

	Raspberry Pi 1 Model A	Raspberry Pi 1 Model A+	Raspberry Pi 1 Model B	Raspberry Pi 1 Model B+	Raspberry Pi 2 Model B	Raspberry Pi 3 Model B	Raspberry Pi Zero
USB 2.0 Ports	1	1	2	4	4	4	1 (Micro-USB)
Ethernet	None	None	10/100 Mbit/s	10/100 Mbit/s	10/100 Mbit/s	10/100 Mbit/s	None
Bluetooth	None	None	None	None	None	4.1	None
WiFi	None	None	None	None	None	802.11n	None
Audio in	I^2S	I^2S	I^2S	I^2S	I^2S	I^2S	I^2S
Audio out	I^2S, analog (3.5 mm jack), digital (HDMI)	I^2S, analog (3.5 mm jack), digital (HDMI)	I^2S, analog (3.5 mm jack), digital (HDMI)	I^2S, analog (3.5 mm jack), digital (HDMI)	I^2S, analog (3.5 mm jack), digital (HDMI)	I^2S, analog (3.5 mm jack), digital (HDMI)	Digital (mini-HDMI), analog GPIO PWM
Video in	CSI Camera Connector	CSI Camera Connector	CSI Camera Connector	CSI Camera Connector	CSI Camera Connector	CSI Camera Connector	None
Video out	HDMI, Composite (RCA)	HDMI, Composite (TRRS)	HDMI, Composite (RCA)	HDMI, Composite (TRRS)	HDMI, Composite (TRRS)	HDMI, Composite (TRRS)	Mini-HDMI, GPIO Composite
External storage	SD	MicroSD	SD	MicroSD	MicroSD	MicroSD	MicroSD

FIGURE 7.10

Comparison of various models of Raspberry Pi.

memory slot, WiFi and Bluetooth module, an audio jack, and USB ports. Due to its very small size it is commonly used in Internet of Things and wearable devices. There are many different Raspberry Pi models. Fig. 7.10 compares them.

As can be seen, the Raspberry Pi 3 model B is the most suitable one for our device as it fulfils all our requirements.

7.4 Analysis

After passing different images into the machine, with different backgrounds and fonts, we checked how many words were misrecognized. This helped us in deriving the accuracy of the OCR module (Table 7.1).

Table 7.1 Accuracy of optical character recognition module with different words.

Image no.	Total words	Misrecognized words	Correctly recognized words
1	283	17	266
2	285	18	267
3	209	27	182
4	247	9	238
5	68	7	61
6	76	9	67
7	70	7	63
8	145	13	132
9	184	11	173
10	226	20	206
11	120	6	114
12	180	14	166
13	270	12	258
14	136	5	131
15	194	10	184
16	78	6	72
17	169	16	153
18	261	19	242
19	57	4	53
20	138	5	133
21	245	12	233
22	142	7	135
23	91	8	83
24	222	14	208
25	65	3	62
Total	4161	308	3853

The data accuracy of the OCR module can be calculated using the following formula:

$$\text{Accuracy (in percentage)} = \frac{\text{Total correctly recognized words}}{\text{Total words}} \times 100$$

The accuracy of the OCR module of Reading Assistant is 92.6%, after preprocessing the image and using the tesseract module.

7.5 Summary

Reading Assistant is an attempt to lend a helping hand to people with various disabilities like blindness, dyslexia etc. helping them to gain greater independent access to leisure, learning and employment. Currently it can be used to read printed text from various sources like newspapers, books, pamphlets etc. It includes the preprocessing of captured images, making its text recognition more efficient and reliable and uses gTTs, that generates an audio file that will be stored for later use.

Even though this project has exhaustively explored every perspective, there is always a scope of innovation and hence enhancements in future work. This device can be further updated, by training it for handwritten characters to widen the scope of serving from printed text to handwritten text. Also, integration of different native and local language translation for the interface will be a notable enhancement since being able to hear the text in one's native language will make it even more enjoyable and comforting.

References

[1] R. Kodali, G. Swamy, L. Boppana, An implementation of IoT for healthcare, Recent Advances In Intelligent Computational Systems (RAICS), IEEE, 2015.

[2] M.A.L. Ghafri, The challenges that visually-impaired students at Sultaqn Qaboos University face in learning english, in: 3 Rd, Global Summit On Education GSE, Malaysia, 2015. 9–10 March.

[3] C. Wakefield, J. Homewood, A. Taylor, Cognitive compensations for blindness in children: an investigation using odour naming, Perception 33 (2004) 429–442.

[4] R. Shilkrot, J. Huber, M. Wong, P. Maes, S. Nanayakkara, FingerReader: a wearable device to explore printed text on the go, in: Conference on Human Factors in Computing Systems, Association for Computing Machinery (ACM), South Korea, 2015. 18–23 April.

[5] R. Shilkrot, P. Maes, J. Huber, S. Nanayakkara, FingerReader: A Wearable Device to Support Text Reading on the Go, in: Extended Abstracts on Human Factors in Computing Systems, Association for Computing Machinery (ACM), Canada, 2014. April 26-May 1.

[6] S. Nanayakkara, R. Shilkrot, K. Yeo, P. Maes, EyeRing: a Finger-worn input device for seamless interactions with our surroundings, in: 4th Augmented Human International Conference, Association for Computing Machinery (ACM), Germany, 2013. 7–8 March.
[7] R. Netoa, N. Fonsecaa, Camera reading for blind people, in: International Conference on Health and Social Care Information Systems and Technologies, Elsevier Ltd., Portugal, 2014. 15–17 October.
[8] M. Bihina Bihina, The phone reader, Undergraduate Thesis, Rhodes University, 2012.
[9] C. Mizan, T. Chakraborty, S. Karmakar, Text Recognition using image processing, IJARCS 8 (2017).
[10] A. Shinde, D. Chougule, Text pre-processing and text segmentation for OCR, IJCSET 2 (2012).
[11] A. Kumar, P. Ahuja, R. Seth, Text extraction and recognition from an image using image processing in Matlab, in: Conference on Advances in Communication and Control Systems, Atlantis Press, 2013.
[12] L. Eikvil, OCR Optical Character Recognition, 1993.
[13] D. Sasirekha, E. Chandra, Text to speech: a simple tutorial, IJSCE 2 (2012).
[14] H. Htun, T. Zin, H. Tun, Text to speech conversion using different speech synthesis, IJSTR 4 (2015).

CHAPTER

8 Toward secure and privacy-preserving WIBSN-based health monitoring applications

Sarra Berrahal[1] **and Noureddine Boudriga**[2]

[1]*Communications Networks and Security Research Laboratory, University of Carthage, Tunisia*

[2]*Computer Science Department, University of Western Cape, Belleville, Cape Town, South Africa*

Chapter Outline

8.1 Introduction

The growing populations of elderly people and patients diagnosed with chronic diseases, the accelerated rate of discovery and spread of new epidemics, the large

Wearable and Implantable Medical Devices. DOI: https://doi.org/10.1016/B978-0-12-815369-7.00008-2

workload of medical staff in emergency departments, and the significantly increasing costs of healthcare services are emerging factors that have led to the unsustainability and inefficiency of healthcare systems in their current form [1]. On the other hand, harnessing the opportunities presentedby the development of miniaturized, smart, and multifunctional sensor devices may provide several technological and commercial benefits to the healthcare industry. These devices can be unobtrusively attached to (or implanted into) the human body, worn by patients or medical practitioners as accessories, or can be part of clothing items and medical equipment [2]. These sensor nodes can be easily networked in the form of a wearable and implanted body sensor network (WIBSN) that allows measurement of various physiological parameters of the patient (e.g., temperature, electrocardiogram (ECG), blood oxygen level), surveillance of the patient's environment, tracking of the patient's motion or behavior. Subsequently, the WIBSN utilizes existing wireless communication protocols to transmit these data to a remote destination for additional processing [3,4]. According to a report published by Grand View Research, Inc., the market of disposable medical sensors is expected to rise more than 12.3 billion dollars by 2025.

Wearable and integrated medical systems have reshaped medical practices through provision of advanced e-healthcare services. Design of an effective WIBSN-based health monitoring system requires careful consideration of a set of practical challenges including: (1) the limited communication and processing capabilities of sensors, (2) the risk of security attacks and confidentiality breaches, (3) the mobility of the monitored patients, (4) the scarcity of wireless network resources, and (5) the lack of effective storage solutions for large amounts of health data. Among these challenges, application security and data confidentiality concerns are paramount due to considerations discussed elsewhere [1]. First, the open nature of the wireless communication medium of WIBSNs and their deployment over unfamiliar and dynamic environments, risks compromise of patient privacy due to a wide range of brute force attacks such as eavesdropping, jamming, false message injection, and denial of service (DoS). In addition, the collected medical data may be shared with various healthcare providers including doctors, paramedical staff, biologists, health research institutes, emergency agencies, and authorized private companies (e.g., clinics, pharmaceutical firms, and insurance companies) to enable a collaborative decision making process. This data sharing creates opportunities where, for example, an adversary (e.g., insurance agent) who is able to illegitimately sensitive data (e.g., therapy specifications, treatment history, etc.) may publicize the data to take revenge on a politician or celebrity, or cause life-threatening risks to the patient by modifying therapy settings, or to rend untrusted the medical institution. Therefore, a security lapse in the healthcare sector not only has serious impacts on the reputation and finances of the targeted healthcare department, but can also lead to dramatic and even lethal consequences for patients depending on the type of attack and the nature of the disclosed data [3,5].

To minimize the impact of security and privacy issues on patients as well as healthcare organizations, privacy rules has been developed by several lawmaking bodies. These rules include The Health Information Technology for Economic and Clinical Health Act (HITECH), the Health Insurance Portability and Accountability Act (HIPAA, the most widely used regulation in the United States), and CEN/ISO EN13606-Part IV (the most widely used regulation in Europe). These rules are generally intended to outline the circumstances under which patients' confidential records may be accessed or disclosed [5,6]. Several approaches for preserving the security and privacy of WIBSN have beendiscussed in the literature and highlighted various techniques including: (1) proximity-based access control techniques that use either ultrasonic distance bounding or energy-aware solutions that aim to minimize the energy overhead, (2) biometrics-based privacy preservation mechanisms that use unique biological properties (e.g., heart signals, retina scan, and fingerprinting) for authentication, and (3) external (wearable) equipment to which the medical device delegates its security. In addition to the advantages provided by these techniques, they also exhibit several limitations including: high energy overhead introduced by the authentication and encryption protocols, fixed credentials used during authentication, inability to effectively avoid DoS attacks, and inability to secure access to \implanted or worn medical devices during emergency situations. Such limitations must be identified and discussed in order to properly address the intrinsic characteristics of wearable and implanted devices.

The main objective of this chapter is to give a thorough analysis of the adoption of WIBSN in the healthcare sector by identifying possible security and privacy threats and describing the associated countermeasures. More precisely, the main topics that will be discussed in this chapter are: (1) describing a typical architecture of a WIBSN-based health monitoring system by identifying the various components that are included in the end-to-end surveillance scenario and by describing how the collected data is communicated between components, (2) presenting WIBSN-based healthcare solutions available in the market, (3) highlighting the primary emerging security and privacy requirements that must be addressed to produce robust WIBSN-based healthcare solutions that fit the newest advances in medical and technological fields, (4) outlining the major security attacks and vulnerabilities that could threaten patients' privacy and lives while identifying countermeasures appropriate for each attack, and (5) reviewing the existing security and privacy preservation schemes that have been recently proposed for health monitoring applications based on the use of WIBSNs and discussing the advantages and limitations of each solution.

The remaining part of this chapter is organized as follows. Section 8.2 describes the motivation of the current chapter and it aims to link the topic covered by the current chapter with the scope of the book. Section 8.3 presents a typical architecture of a healthcare system that integrates wearable and implanted sensing technologies and cloud computing to deliver e-health services. The main health monitoring applications targeted by WISBN are also reported in the same section. Section 8.4 covers various types of attacks that could endanger patients' privacy and safety. A set of imperative requirements are elaborated in Section 8.5. Section 8.6 focuses on countermeasures against the previously described attacks through the description of a set of privacy preservation techniques and mechanisms. Section 8.7 contains analysis and comparison of the described techniques. Section 8.8 outlines challenges and future directions that should be carefully studied in current WIBSN-based healthcare solutions. Finally, Section 8.9 concludes the chapter.

8.2 The chapter's motivation

The present chapter deals with several topics addressed by the current book. Principally, it focuses on the topic of security and privacy in healthcare WIBSN. Indeed, the emergence of advanced wearable and implanted devices (e.g., smart watches, smart retina, implantable medical devices (IMDs), etc.) has facilitated high performance health monitoring. However, WIBSNs are considered as double edged systems. Although they facilitate remote monitoring of patients' health statuses, they can also risk patients' lives if inadequate security measures are taken.

In Fig. 8.1, we illustrate several types and prototypes of wearable and integrated medical devices that are already proposed in the market to treat various diseases such as stroke, heart arrhythmia, diabetes, and epilepsy. These devices can be controlled by a programmer (or a personal server) via wireless communication to remotely monitor the patient's vital signs, check the device safety, and allow doctors to adjust and modify therapy [7]. For instance, analysis of heart electrical activity using a pacemaker (i.e., a surgically implanted device) will

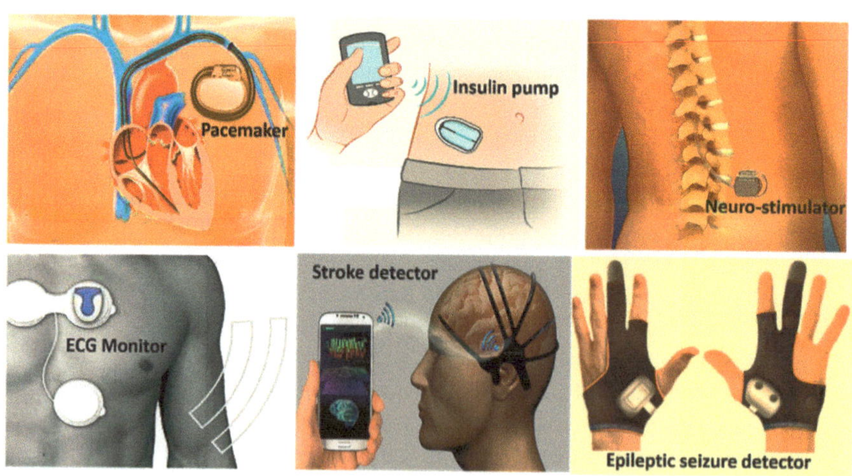

FIGURE 8.1

An illustration of some wearable and implantable medical devices.

facilitate the treatment of heart arrhythmia such as tachycardia and bradycardia. Brain activity may be recorded in order to detect the occurrence of diseases such as Alzheimer and the early signs of epileptic seizures or stroke. Using the collected data, a neurostimulation device appropriately stimulates the brain either directly in the deep brain or from the internal surface of the skull [8,9]. Vital-Connect Inc. (Campbell, CA, United States) is a startup company that has proposed a sensing device that encapsulates eight unique sensors within one device called single-lead-ECG [9]. This device is commercialized in the form of a patch that enables the monitoring of different physiological parameters including posture, body temperature, as well as respiratory and heart rates. In 2015, a group of engineers from Samsung's Creative Lab (C-Lab) introduced a prototype of a stroke detection device that enables the monitoring of electrical brain impulses using a wearable sensor suite. This device interacts wirelessly with a tablet or smartphone in order to warn the user of an oncoming stroke.

Several critical security and privacy issues must be properly addressed before the deployment of WIBSN-based solutions because of the sensitive and confidential information collected by the various sensors, which can be attached to or worn by the patient and may be deployed either inside the medical environment or outside of it (e.g., residential areas [10]). The collected data is exchanged between various healthcare providers and others to improve medical cooperation and optimize the decision making process. In addition, as WIBSN may be deployed over unfamiliar and dynamic environments, several brute force attacks (e.g., eavesdropping, jamming, and data modification) could be executed. These attacks may compromise patients' privacy and could be lead to damage to the reputation and finances of the healthcare department. Life-threatening consequences for patients could also result. Accordingly, the main objective of this chapter is to thoroughly analyze the security and privacy issues that have emerged with the adoption of WIBSN in the healthcare sector.

8.3 WIBSN-based healthcare system

Small size and low cost of miniaturized wearable devices including sensors and actuators are key factors in the use of WIBSNs in a wide range of disciplines related to biology and medicine. In this section we will describe a typical architecture of WIBSN-based healthcare systems through the identification of several components required to build an end-to-end healthcare service. We will also describe the different communication scenarios that could exist between them.

8.3.1 The system architecture

To describe the main components and monitoring services provided by such WIBSNs, we illustrate in Fig. 8.1 a conceptual representation of a WIBSN-based architecture of a healthcare system. The WIBSN is composed of multiple components that work together in order to collect, process, and transmit patients' information to an authorized remote or local medical department. The presented architecture will enable us to simplify the description of the major security issues (see Section 8.4) that could threaten the success of the healthcare system.

The illustrated system incorporates a number of patients (or medical practitioners). Each is represented in a WIBSN formed by wearable as well as integrated sensors in charge of regular collection of various physiological and environmental measurements and administration of medicine. Sensor devices used in healthcare are designed such that they can be either implanted (i.e., surgically) under the skin (e.g., neurostimulators, retina prostheses, pacemakers, glucometers, cardiac defibrillators, and insulin pumps) or worn on the body (e.g., smart watches, smart badges, glasses, bracelets, and ECG sensors). These sensors allow greater freedom of movement for patients, while allowing the collection of data on his/her vital signs such as body temperature, heart rate, blood saturation, and ECG. Some devices such as the insulin pump play the role of an actuator node, which is in charge of performing critical actions such as the administration of the appropriate dose of drugs (i.e., insulin in the case of a diabetic patient) [11]. Such action is taken based on data (e.g., glucose level measurement) received either from sensors in the same WIBSN, or from the user via the body node coordinator (BNC) [12]. The latter is assimilated to a personal server that orchestrates the interactions among all the sensors of the WIBSN and between the WIBSN and the surrounding environment.

Compared to sensors, the BNC is a powerful device endowed with enhanced communication and processing capabilities. It is responsible for collecting sensor readings, preprocessing them in real time before referral, storing them temporarily (if necessary), and then transmitting them wirelessly using any available ambient networks (WiFi, satellite, 3G/4G, etc.) to the remote healthcare processing unit. The collected data is forwarded by the BNC toward sink nodes (e.g., access point, gateway node, or mobile node) to their destination. Commonly, the BNC can be a personal digital assistant, a smart mobile phone, or any connected device. Due to wireless resource scarcity (short-range communication and limited power consumption) and problems linked to the dynamic environment where WIBSNs are evolving (e.g., variable path loss), it is critical to improve communication reliability and minimize power consumption, in order to enhance the network's lifetime. In this context, multihop cooperative communication approaches have been proposed and implemented in several solutions discussed in the literature ([2,13,14], and [15]). In these implementations a single mobile WIBSN or a set of mobile WIBSNs may serve as an intermediate node or network to relay the data collected from one or several BNC(s) to their destination.

The emergence of new technological paradigms such as cloud computing has offered new opportunities for WIBSNs by enabling expansion of their sensing objectives and monitoring functionalities [8,16]. In fact, current sensor-based healthcare applications have evolved from being departmental solutions delivering a defined healthcare service to a specific type of patients (e.g., ECG monitoring in patients with cardiac disease) to large-scale solutions that encompass several interconnected systems and provide several types of services to a large community of users. Though WIBSNs have limited communication, processing, and storage resources, these limitations are offset by benefits including, but not limited to, provision of on-demand access to a shared pool of computing resources and large storage services, support of voluminous and heterogeneous health record offloading from many connected sensors and medical devices (e.g., sensors readings, radiology images, and genomic data), and provision of real time and cost effective health monitoring services. Accordingly, collected records are transmitted to the cloud domain where they are postprocessed and stored in distributed cloud storage systems and accessed by several medical parties and departments with different specialties, requirements, and profiles. For example, doctors require access to all kinds and levels of patient data. Doctors are assisted by nurses and other paramedic staff that require access to patient personal information and sensor readings. The administrative staff of a medical department are authorized to access patient personal and insurance-related information to perform billing operation. Patient families may receive alerts about patient health status. Pharmacists may only need access to drugs prescribed to the patient. Patients are data owners and are therefore authorized by default to full access to his/her data stored in medical databases.

The described WIBSN-based architecture could be vulnerable to several types of security threats during data collection using resource-constrained sensors, data transmission to authorized healthcare providers using wireless environment and web portal, as well as during data storage and processing over distributed storage systems. In fact, during the transmission of health records from sensors worn or implanted on/in the human body, attackers could potentially execute illegitimate energy-hungry actions from several unauthorized sources in order to manipulate the collected vital signs. In addition, criminal intruders could execute DoS attacks to prevent the coordinator node of the WIBSN (e.g., a smartphone) or to the remote medical server from correctly receiving the collected sensitive health data. Manipulating the collected vital signs and preventing their proper transmission can result in serious issues such as the misunderstanding of the collected medical report (i.e., some records are missed), the generation of false alerts, and the non-detection of a critical emergency situation (e.g., a heart attack). A set of common attacks that could be conducted on wearable and implanted sensors will be identified in the next section. However, we will first describe the communication architecture required to ensure end-to-end health data delivery.

8.3.2 A three-layered communication architecture

In general, three different layers of communications are supported by the previously described architecture; namely, the intra-WIBSN communications, the inter-WIBSN communications, and the extra-WIBSN communications. These different levels of communication are illustrated in Figs. 8.2 and 8.3.

- *Intra-WIBSN communications* refers to the short-range communications (i.e., about 2 or 3 meters around the human body) that occur among the different implanted or wearable sensors forming the WIBSN, or between sensors and the BNC [17]. Indeed, the several sensors in a WIBSN are elementary devices that are able to communicate with each other through wireless or wired media and send their readings to the BNC. However, problems arise due to movement of the human body, the location of the sensors, and the fading of wireless communications (e.g., reflection, diffraction, shadowing, and energy absorption by variable body postures and clothing). In order to overcome these in-body communication problems, multihop topology is employed by allowing nodes to act as relays to transmit the collected data to the BCU [14]. Well known communication standards for ensuring intra-WBAN communications include: radio-frequency identification (RFID), ANT, human body communication, ultra wide band (UWB), Zigbee, and Bluetooth [17].
- *Inter-WBAN communications* refers to the communications between two or more WIBSNs via their BNCs in order to extend the communication range. As previously mentioned, in every WIBSN, the BNC node plays the role of an intermediate device between the WIBSN and the surrounding environment. The BNC is responsible for collecting data from elementary sensor nodes associated to the same network, acting as a relay node for other WIBSNs in

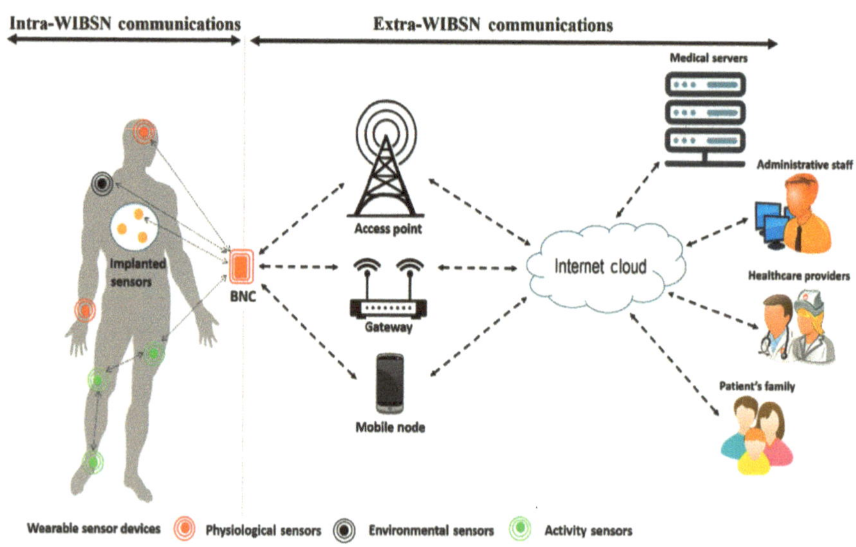

FIGURE 8.2

An illustration of a typical wearable and integrated body sensor network (WIBSN).

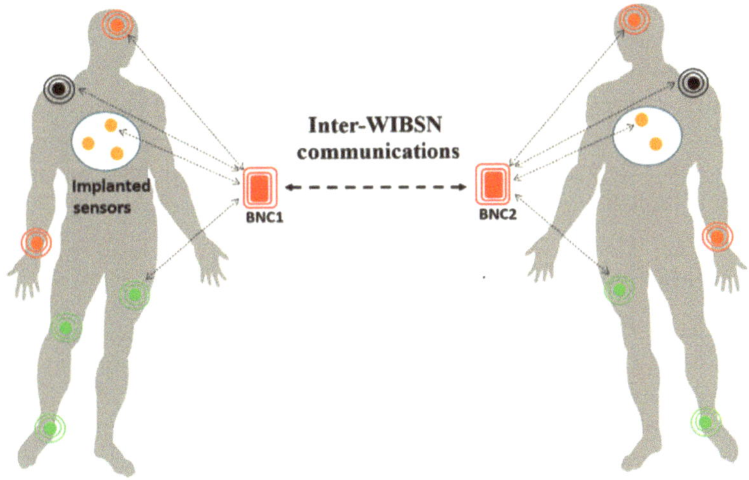

FIGURE 8.3

An illustration of inter-WIBSN communication.

close vicinity, and then forwarding the received data to the remote medical center, through any ambient network (e.g., WiFi, 3G, 4G, etc.) [2]. In fact, due to users' mobility, the limited energy and communication resources of sensors, and the incomplete availability of communication infrastructure in some environments such as rural areas, inter-WIBSN cooperative communications are needed in order to achieve reliable data transmission in case of out-of-coverage or unavailable network infrastructures. Nonetheless, enhancing data reliability should takes into consideration security issues that could emerge. Common wireless standards used for inter-WIBSN communications are RFID, UWB, Zigbee, Bluetooth, and Bluetooth Low Energy [9,15]. Nevertheless, it is also possible to use other wireless standards such as WiFi or cellular data services (e.g., GPRS, 3G, or 4G). The selection of one of these standards depends generally on the objectives and range requirements of the conceived healthcare application.

- *Extra-WBAN communications* refers to the long-range communications that connect the WIBSNs to the cloud environment via several standards such as WiFi, cellular data services (GPRS, 3G, or 4G), Zarlink and RuBee. In the cloud environment, storage and computing resources are organized in a virtualized pool and used by several healthcare providers. Indeed, medical departments use cloud resources in order to run applications and deliver advanced remote healthcare services. The data collected from the networks of WIBSNs will be stored in distributed medical databases where they will be postprocessed (e.g., filtered, aggregated, and analyzed) by authorized parties (e.g., doctors, paramedic staff, scientists, etc.) and inform medical decisions. In addition, according to the application requirements, the health data may be accessed not only by doctors or caregivers' staff, but also by pharmacists, administrative staff, insurers, patients' families, and the patient himself or herself. For example, healthcare providers are able to diagnose a health problem by interrogating health records stored in the cloud databases or by directly retrieving data from a WIBSN. In the case of emergency detection such as exceeding a given threshold or identification of an infection, medical action is taken accordingly (Fig. 8.2).

8.3.3 Health monitoring applications

Applications of WIBSNs have initially focused on physical medicine and rehabilitation, with the goal of improving quality of lifeby enabling continuous monitoring and processing of vital parameters of patients suffering from chronic diseases (e.g., cardiac failure, diabetes, renal disease, and asthma). The first type of sensors used by medical research teams were accelerometers or a combination of accelerometers and electromyography sensors that measure human behaviors that vary overtime (e.g., movement and muscle activity patterns associated with a given set of identified physiological tasks). The monitoring applications are generally organized into two types: medical and nonmedical applications. Here, we report the primary medical monitoring purposes of WIBSNs.

Remote medical monitoring. Wearable and implanted sensors are integrated in the WIBSN to measure several physiological and vital parameters of the patient in residential [18] or medical environments. Wearable devices such as temperature sensors, respiration monitors, pulse oximeters, pH monitors, and glucose sensors can be worn by the patient or attached to the body surface. IMDs are surgically implanted inside the patient's body. Collected vital signals are transmitted to a remote medical center, which should provide real time feedback and remote configuration of the medical device. Such an application facilitates doctors' tracking of patients at any location and under everyday conditions without constraining their daily activities.

Disease-specific monitoring applications. The goal of such systems is to enable monitoring of chronic diseases (e.g., cardiovascular disease [12,19], Parkinson's disease, Alzheimer's disease, epilepsy [11], and diabetes), cancer detection, and stress monitoring. WIBSNs can detect irregular physiological activities of the patient's body and prevent death through preemptive prescription of appropriate treatment. In addition, through tracking of a disease symptoms, some epidemics could be easily detected and their evolution could be tracked in real time [13].

Rehabilitation and aging individuals monitoring applications. Such applications will provide remote monitoring to chronically ill patients or elderly people in their residential environments without constraining their daily activities. These application will additionally minimize the high costs of hospitalization and reduce frequent hospital visits and medical workload [18]. To this end, a wide range of implanted and wearable sensors for event detection, daily activity tracking, medical coordination, and reporting have been proposed in the market. For example, after being dismissed from the hospital, a patient can be equipped with a WIBSN-based medical system programmed to continuously monitor the patient's health status. The analysis of the collected health records will facilitate the surveillance of health complications and the reestablishment of their vital functions.

8.4 Primary attacks targeting healthcare applications

Because each communication layer has its own characteristics, the identified vulnerabilities, the impact of conducted attacks, and the corresponding countermeasures will vary from one communication layer to another. In general, attacks are guided by two primary motivations:

1. The first motivation is to affect data privacy through the disclosure of confidential and highly sensitive information (e.g., patient identity, location, profession, sexual orientation, etc.) via passive attacks such as eavesdropping and traffic analysis/monitoring attacks. Formally, a passive attacker aims to steal exchanged data without interfering with the radio communication.
2. The second motivation is to compromise and damage the healthcare system using malicious attacks such as jamming or DoS attacks in order to deplete the network's resources to cause malfunctioning or inoperability. More explicitly, the attacker aims to disrupt network availability by cutting off communication links between the connected nodes, modifying and tampering with the collected data, or even by injecting fake data.

While some attacks have a single security motivation, several types of attacks are motivated by the desire to steal private and confidential data in order to compromise a system's availability and integrity. The potential attackers can be either internal malicious employees (e.g., administrative or medical staff of the medical department) abusing their privileges, or external attackers that have successfully exploited the system's vulnerabilities [3,20,21]. In this section, we provide an in depth analysis of the major attacks that could be conducted on a WIBSN-based healthcare system and we will try to identify several countermeasures that could be taken in order to minimize the scope of these attacks.

8.4.1 Eavesdropping on radio communications among sensors

An eavesdropping attack also known as spoofing or sniffing is an attack conducted against health data privacy and confidentiality, even if encryption mechanisms are considered during transmission. Eavesdropping refers to the act of illegally intercepting private conversations between two communicating entities in order to steal sensitive data that should be kept secure during transmission. Snooping on exchanged data between wireless devices may reveal information related to the duration and time of a communication, the identities of the communicating parties and their location, and the complete shape of the exchanged traffic. Therefore, if the captured data is not encrypted, its content can be easily analyzed to extract sensitive information related to patient health statuses or confidential medical facts (e.g., physiological parameters, therapy and treatment specifications, visit history, sexual orientation, and psychiatric profile) or network configuration and access control information. Although after an eavesdropping attack the integrity of the data remains, its privacy is compromised. In fact, the eavesdropper maybe able to process the captured information in order to conduct other types of attacks that place at risk system operability. Similarly to wireless communication eavesdropping, attacks such as traffic analysis and traffic monitoring can be also conducted in order to infer information about the wireless communication [4].

To illustrate the impact of an eavesdropping attack on medical devices, we consider the two following examples: (1) Using an unauthenticated programmer such as a Universal Software Radio Peripheral, a malicious person can easily conduct an eavesdropping attack on an IMD by overhearing the exchanged messages between the communicating entities. These entities are the IMD and the programmer device, or the wireless medical sensors and a medical implant communication system (MICS) transceiver in the IMD. Since the exchanged data are generally not encrypted, they can be analyzed to infer sensitive information related to patient physiological data, therapy specifications, treatment history, the identification number of the IMD, as well as the personal identification number of the IMD used to exchange data in cleartext [7]. The collected information can be exploited to perform unlawful actions, such as tampering with therapy settings, that could put patient health at risk. (2) A fingerprint and timing-based snooping attack can be executed in order to snoop on radio transmissions among sensors in residential and healthcare environments. Clandestine actions such as inter-sensor and extra-sensor communication tracking can be performed in order to collect the timestamps and fingerprints of the exchanged messages. The stolen fingerprints are then used in multiinference phases in order to deduce the sensors' locations and types. This information is used as an input in order to track private daily activities (e.g., cooking time, showering time, sleeping time, number of residents, and number of rooms) and infer health-related information.

8.4.2 Denial of service: attacks against system availability and integrity

A crucial requirement for the provision of a trustworthy and effective WIBSN-based healthcare system is to keep the implanted and wearable devices available for their intended use throughout the monitoring process. However, as previously discussed, such a system is vulnerable to potentially dangerous and illegitimate actions that aim to lower the performance of the deployed network through the execution of resource-hungry DoS attacks that overload the scarce network resources. For example, the execution of repetitive authentication requests will force a sensor to act on every request until its battery is completely depleted and consequently it become inoperable [22]. DoS attacks can be conducted either against system availability or against data integrity. Availability threatening attacks include, but are not limited to, jamming, data collision, data flooding, unfairness attacks, and desynchronization attacks. Attacks against service integrity aim to conduct illegitimate actions such as data tampering, false data injection, and data removal in order to affect the decision making process.

Here, jamming, false data injection, and Sybil attacks are described:

Conducting jamming attacks on WIBSN: Ajammer aims to disturb signal transmission through the creation of electromagnetic interferences in the WIBSN operational frequencies [3]. The main goal of such an attack is to isolate the wearable and implanted medical sensors of the WIBSN by cutting off the wireless communication links among them and their corresponding destination (e.g., the BNC, the gateway nodes, and the WIBSNs in their close vicinity), and subsequently prevent the transmission and reception of legitimate medical information such as notification of vital signs degradation and appropriate treatment delivery in response. The jamming attack can disturb the entire health monitoring process by randomly distributing several jamming nodes in the network. The consequences of such an attack can be lethal in the case of critical and time sensitive communications. For example, a malicious user may interfere with the communication channel between the IMD and the programmer in order to block traffic exchange between them and prevent reception of the required configuration and delivery of logged events. The jammer can also target the wireless communication between the IMD and other medical sensors in close vicinity in order to prevent the IMD from collecting critical events detected by these sensors. These two scenarios will prevent delivery of appropriate therapy and swift reaction in case of emergency.

Injecting false data on WIBSN sensors: Wearable and implanted medical sensors are not tamper resistant and can therefore be easily captured and compromised by an attacker that was able to gain illegal access to the network. In this case, the attacker intercepts exchanged private information in order to alter the communicating nodes and inject false sensing reports. This creats a number of attacks (e.g., black hole attacks and flooding attacks) to consequently undermine the capacity of the healthcare system in the detection of valid medical events. Therefore, both system integrity and availability can be easily compromised by a false data injection attack. In fact, the delivery of fake sensing reports to the remote medical center leads to the creation of false alarms that waste network resources (e.g., bandwidth and energy). For example, an attacker can inject false configuration settings in the IMD in order to disable therapies (e.g., stop the appropriate injection of insulin) and even deliver a physiological shock (e.g., inject a large dose of insulin) that may kill the patient. Other unauthorized commands can be executed such as tampering with IMD firmware (i.e., the actuator will react unpredictably), incorrectly configuring the IMD clock (i.e., the creation of invalid and ambiguous timestamps in the IMD event log), removing records stored in the IMD, or disabling IMD functions. Another attack that should be carefully addressed is clogging attack, a serious degradation of service attack where bogus encrypted messages are transmitted by the attacker. Before rejection of these messages, they are decrypted by the IMD. The repeated transmission of clogging messages unavoidably leads to depletion of the IMD battery.

Stealing the identity of sensors nodes using Sybil attack: An adversary is able to conduct a destructive Sybil attack against the integrity of wearable and implanted medical sensors by using forged identities to illegally access the network. A Sybil node can be described as a misbehaving node that simultaneously claims multiple identities, which causes confusion in the monitoring system. A Sybil attack can create redundancy problems during data transit and routing, as well as degrade data integrity and deplete resources. A fake node can interfere with insulin administration in an implanted insulin pump, which may cause hypoglycemia (i.e., low level of blood glucose) or hyperglycemia (i.e., high level of blood glucose), which may put at risk the diabetic patient's life [11].

8.5 Security and privacy requirements

Common security and privacy requirements of WIBSNs include preservation of content confidentiality, assurance of effective authentication, guarantee of end-to-end data integrity checks, and enhancement of auditing capabilities. However, due to the constrained resources of WIBSNs in terms of limited energy and restricted computational and communication capabilities, as well as security vulnerabilities associated with wireless environments, the security specifications proposed for other networks are not applicable to WIBSNs. Here we discusssome security and privacy requirements that should be carefully considered.

The assurance of health data confidentiality is a fundamental security service for any healthcare system in order to ensure that data is protected against illegitimate disclosure to unauthorized parties [3]. Confidentiality is the matter of providing access to private health records to authorized and correctly authenticated entities. As stated before, WIBSNs may be vulnerable to brute force attacks that may compromise the confidentiality of the communication environment and consequently impact the monitoring system as a whole (i.e., sensors, communication nodes, and storage devices). For example, an attacker can easily eavesdrop on radio communication between system nodes in order to disclose sensitive data to unauthorized users. Therefore, confidentiality must be maintained during the entire health monitoring process (i.e., data collection, transfer, storage, and retrieval) by encrypting the collected data. In addition, confidentiality must consider governmental law and regulation standards such as HIPAA in the United States and CEN/ISO EN13606-Part IV in Europe that require protection of data confidentiality. Although data confidentiality improves system privacy by ensuring that the data sources are verifiable, it does not guarantee that the exchanged data has not been interfered with or tampered with during transmission.

The guarantee of end-to-end health data integrity is related to the protection of health data from malicious modifications by ensuring consistency and accuracy of data from the source to the destination. Ensuring end-to-end data integrity checks is crucial in critical applications like healthcare because accurate monitoring of patient health statuses depends onvalues transmitted from the medical sensors. If these values are tampered with, then a patient's life can be put in danger. For example, an attacker that has eavesdropped on the radio communication between a WIBSN and the medical server can perform a data tampering attack on the disclosed information in order to achieve criminal purposes such as the alteration of therapy settings or modification of an issued medical command. Such actions may damage the healthcare system and cause life-threatening consequences for patients.

The assurance of healthcare services availability is among the crucial technical requirements that should be guaranteed. Indeed, healthcare systems must ensure that medical data is easily retrievable anytime and anywhere without service interruptions or degraded performance. In addition, since the availability of a health monitoring system is generally threatened by several types of DoS attacks such as jamming, desynchronization, data tampering, and packet flooding, appropriate countermeasures that take into consideration the characteristics of medical sensors and the importance of healthcare applications should be designed.

The implementation of a secure access mechanism and authentication service makes medical records accessible by various medical parties and departments with different specialties and requirements. In addition, each patient carrying a WIBSN system and participating in the health monitoring process must be authenticated prior to being allowed to access the medical equipment and to participate in the monitoring process [23]. Unlike other collaborative first responder teams that are easily predefined and static, healthcare teams are dynamically formed with distinctly different team members. In general, the process of team formation during a health diagnosis is triggered by the arrival of a patient or the reception of patient health data. The members belonging to the same team may have different authorization levels. For example, while doctors are able to access all types and levels of patient data, administrative staff (e.g., receptionists and ward staff) are only authorized to access personal and insurance-related information for billing operation purposes. Therefore, granting all users the same level of access without considering distinctions insensitivity levels can be problematic. On the other hand, to ensure that the received health records are transmitted from a trusted node, authentication services are needed to check the communication validity.

Freshness protection service ensures that received medical records are recent and that no redundant copies of data that has been previously received exist. In fact, data freshness can be threatened by an attacker that has successfully eavesdropped the network and performed a replay attack by retransmitting the previously collected data. Ensuring privacy-aware services such as confidentiality and authentication is not sufficient to counter replaying attacks, especially in networks with shared communication and processing resources [21]. By attaching a randomly generated number or a time dependent token to the transmitted data, data freshness can be ensured by approving only recent data through the rejection of messages with old time tokens and a randomly generated number that was previously generated.

The provision of real time responsiveness and reliable services: The deployment of WIBSNs is associated with some communication issues such as frequent fluctuations of network resources and communication link unavailability due to nodes mobility and resources scarcity [24]. In addition, due to their constrained resources, communication among sensor nodes is generally error-prone and unreliable. Thus, uploading sensed data to the remote healthcare system becomes a significant bottleneck. Indeed, according to [2], the generated medical traffic can be classified into emergency traffic, real-time traffic, and best-effort traffic, which do not tolerate delays, have stringent delay requirements, and tolerate a bounded time interval, respectively. The evaluation of data related to ordinary monitoring records is generally delay tolerant. In spite of this noteworthy difference, these three types of data share a requirement to be reliably managed within a specified deadline. Otherwise, the collected data may no longer be useful and the related healthcare system may be unable to effectively provide services, thus threatening the life of the patient.

Management of health data privacy while considering data variability and dependability: Different types of dependencies can exist between the health data stored within the distributed storage systems of the cloud. The capability of tracing data dependency in healthcare applications will facilitate the identification of eventual diseases, prevent disease occurrence and propagation, and will help in the provision of customized medical diagnosis. However, if an attacker is able to exploit and trace inter-data dependencies, it will be able to identify patients. In addition, due to the variable and complex nature of the monitored environment as well as the evolving nature of health-related data, time-dependency is considered a key aspect of effective data security. Traditional health data privacy management approaches that assign a static privacy status to medical data may not be appropriate.

In Table 8.1, a classification of the main security and privacy requirements that should be considered by any e-healthcare system is illustrated. Generally, security measures should be applied in three main security levels: The information (sensors) layer, the network layer, and the administrative layer.

1. Security measures at the information level must comply with a set of requirements including, but not limited to: development of effective

Table 8.1 Requirements for securing WIBSN-based healthcare applications.

Requirements	Security level		
	Information	Network	Administrative
Encryption mechanism	✓		
Immunity to message tampering	✓	✓	
Authentication	✓		
Freshness	✓		
Data availability	✓		
Confidentiality and privacy preservation	✓		
Anomaly detection and reaction	✓	✓	✓
Trust and reputation management	✓		✓
Secure nodes localization	✓		
Intrusion detection mechanisms		✓	
Attack prevention mechanisms	✓	✓	✓
Data access control			✓
Accountability			✓
Revocability			✓
Timeliness			✓
Patient anonymity		✓	✓
Immune to activity tracking		✓	✓
Audition services			✓
Secure data transmission		✓	

encryption mechanisms, assurance of system immunity against data tampering attacks, secure authentication of authorized users, guarantee of data freshness to prevent packet replay attacks, maintenance of data and healthcare services availability, implementation of confidentiality and privacy preservation mechanisms, detection of anomalies and deployment of appropriate countermeasures, management of system trust and reputation, and provision of secure localization of the connected nodes.

2. Security measures at the network level are required because the wireless communication medium between sensors is generally vulnerable and prone to various brute force attacks as well as to several environmental issues (e.g., variability of the communication environment, malfunction of medical sensors due to signal shadowing and the presence of obstacles in the surrounding environment, possible destruction of sensors, degradation overtime of communication resources and system conditions, and depletion of sensor energy due to extended operation). Such problems can be resolved by ensuring a set of requirements such as the design of intrusion detection as well as attack prevention mechanisms (e.g., prevention of selective and replay forwarding, prevention of black hole and sinkhole attacks, countering Sybil attacks, and opposing session hijacking and energy depletion attacks).

3. At the administrative level mechanisms that enable the prevention and detection of security breaches must be implemented by the administrator system. Therefore, a well-defined hierarchy-based access control (i.e., taking into account the different parties participating in the healthcare process and their roles and requirements) in combination with strong authentication measures can be beneficial. In addition, security measures that ensure a set of requirements such as data access control, accountability, revocability, timeliness, and immunity against activity tracking should also be considered. As health data could be shared among several healthcare providers and users for additional exploration or for improved decision making, patient anonymity should be respected. Audit services are also important in order to record and track all read/write activities on health data.

8.6 Security awareness and privacy preservation techniques

Several mechanisms for the preservation of privacy and security of sensors-based healthcare applications have been reported in the literature and have highlighted various techniques (e.g., proximity-based access control, key distribution mechanisms, and biometrics) that attempt to minimize the likelihood of successful attacks on WIBSN. A detailed description of these techniques is provided below.

8.6.1 Proximity-based access control mechanism

The proximity-based access control (PBAC) mechanism provides automated access to resources based on calculation of the user's proximity [25,26]. The proximity is not only related to the physical proximity between two communicating devices, but it can be associated to other dimensions including temporal proximity, geospatial proximity, risk proximity, and organizational proximity. Based on the calculated distance, only devices located within an acceptable security range from the medical device are granted access to its resources. This security range is generally defined based on the medical task. For example, critical medical interventions such as configuring therapeutic settings of the implanted/wearable medical device requires a much smaller security range than ordinary remote monitoring tasks [5]. The PBAC-based mechanism has been adopted differently in several publications [16,27,28]. Among these works we describe here two main approaches. The first has been proposed in [28] and uses ultrasonic distance bounding (Ultrasonic-AC). The second technique discussed in [16] is based on zero-power defenses and a piezo-element and tries to minimize energy consumption using lightweight cryptographic protocols [29].

8.6.1.1 Ultrasonic-AC approach

In [28] an Ultrasonic-AC approach is described in order to measure the distance separating the IMD from the programmer using a device pairing protocol that combines proximity verification and credentials calculation. A security token is carried by the patient and shares a secret key with the IMD. In ordinary monitoring operations, this token is used to allow the programmer placed (by the caregiver) within a defined security range from the IMD to access to the IMD. During emergencies the token is unavailable; therefore a pairing protocol that implements the Diffie-Hellman (DH) algorithm is used. DH facilitates the establishment of a shared secret among two communicating entities. Using the DH pairing tool, the IMD generates a temporary shared key, which will be used to initiate the sharing of the encrypted medical records with programmers in its close proximity [12]. For the sake of simplicity the exchanging of DH keys can be implemented using the multiplicative group of integers modulo p (i.e., p is a prime and g refers to a primitive root modulo p). Accordingly, the resulting shared secret can take any value from the interval $[1, p-1]$. To describe the DH protocol let us consider two communicating parties, in our case the programmer and the IMD. The programmer selects an integer a and transmits $A = g^a mod\, p$ to the IMD, which in turn picks an integer b and publishes $B = g^b mod\ p$. The programmer and the IMD compute $K_A = B^a mod\ p$ and $K_B = A^b mod\ p$, respectively. The same secret K is currently shared between the two communicating parties since $g^{ab} mod\ p = g^{ba} mod\, p$ and consequently $K{:} = K_A = K_B$. Once this shared secret is computed, the programmer and the IMD can use it as an encryption key to secure their communications.

The pairing protocol proposed in [28] is illustrated in Fig. 8.4 and is initiated on the programmer side by the selection of a secret component p, which is used in order to generate a nonce N_p. Then the programmer computes the DH public part, which is g^p and sends an authentication request to the IMD. The latter selects in turn a nonce N_v, replies with a single bit of N_v, and saves the transmission time t_1. The programmer initiates with the IMD a bit-by-bit exchange of the generated nonce. At the reception of the first bit, the IMD, the programmer performs a xor function on N_v and a single bit of g^p and sends back the result as a sound message. The programmer uses the received messages in order to calculate the distance separating the two devices as follows: $d = (t_{i+1} - t_i).v_s$, where t_i, t_{i+1}, and v_s stands for the first bit reception instant, last bit reception instant, and the sound speed, respectively. The programmer then compares the computed distance with a threshold distance d_{thresh} referring to the safety proximity. So, if $d < d_{thresh}$, the programmer is located in the close proximity of the IMD and continues running the pairing protocol. Otherwise the IMD terminates the session. The programmer similarly executes a proximity verification of the IMD. To this end, it selects a prime v, computes the corresponding g^v, and achieves a bit-by-bit exchange. At the end of the communication process the programmer sends a message authentication code $MAC(N_v, N_p)$ using the two nonce N_v and N_p and a key $(k = g^{v^*p})$ that has been successfully established.

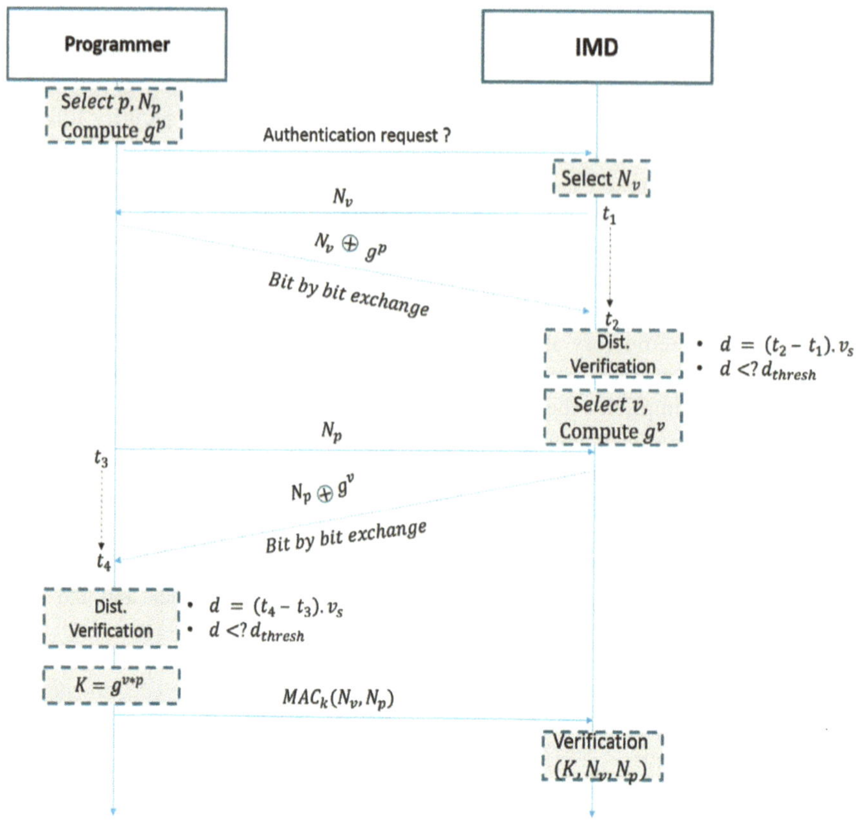

FIGURE 8.4

An illustration of the device pairing protocol using DH.

8.6.1.2 Energy-aware proximity-based access control techniques

Energy-aware PBAC techniques referred to as zero-power defenses have been proposed in the literature to harvest radio frequency (RF) energy from an external device (e.g., piezo-element) instead of drawing from its own energy. For example, in [16] a solution that integrates three zero-power defenses (i.e., zero-power notification, zero-power authentication, and sensible key) and uses a piezo-element implanted in the patient's body is proposed. The piezo-element generates a zero-power notification that audibly warns the patient about the occurrence of both internal and external security-sensitive events without draining battery power. To authenticate requests from an external device programmer, a zero-power authentication approach is used. This appraoch harvests RF energy to power a cryptographically strong protocol to manage the communication between an implantable cardioverter defibrillator programmer and a Wireless Identification Sensing Platform (WISP) RFID tag.

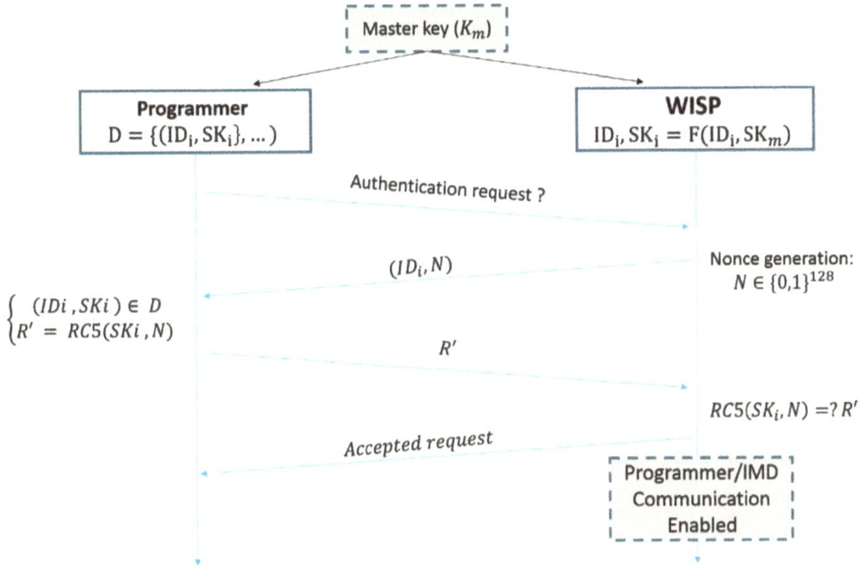

FIGURE 8.5

Communication establishment between an implantable cardioverter defibrillator (ICD) programmer and the WISP RFID Tag.

The authentication protocol is described in Fig. 8.4. A simple challenge response protocol based on the RC5−32/12/16 communication protocol. In addition, a zero-power authentication device (i.e., a WISP RFID tag) is used and a master key K_m is known by any programmers. A unique serial number or identity ID_i is attached to each IMD and a cryptographically pseudorandom function (e.g., AES) is used to compute an IMD-specific key: $K = F(ID_i, K_m)$ [16]. The master key value K_m should be kept secure in the programmers. The authentication protocol begins by the reception of an authentication request from the programmer. The WISP answers with a message containing a nonce N and its identification number ID_i. Using the received WISP identity, the programmer computes the IMD-specific key $K = F(ID_i, K_m)$ and then returns to the WISP the response: $R' = RC5(K, N)$. In turn, the WISP computes the response R and checks the validity of the value received from the programmer: $RC5(SKi, N) = ?R'$. If the challenge is successfully passed, then the programmer will be authorized to access the IMD and the WISP sets a general-purpose I/O (GPIO) high. The zero-power notification and zero-power authentication are combined to allow patients equipped with IMDs to physically sense the exchanged acoustic key (Fig. 8.5).

8.6.2 Biometrics-based privacy preserving mechanisms

Biometrics have been widely implemented as cost-effective authentication solution for secure implanted medical sensors [12,30−35]. In fact, compared to generic cryptosystems, biometrics enable higher security levels with less memory and computation requirements (i.e., they can generate strong randomized keys with a higher level of entropy used for data encryption purposes), which makes them suitable for securing wearable and implanted medical systems. Biometrics refers to behavioral or physiological measurable properties that allow unique identification of an individual. In general, an ideal biometric property should satisfy the following characteristics: (1) Universality: the biometric attribute should be possessed by any living human subject; (2) Permanence: constant over a reasonable period of time and not be subject to significant alteration; (3) Uniqueness: have sufficient unique features that differentiate an individual from any other; (4) Robustness: invulnerable and extremely difficult to be reproduced or mimed by malicious actors [30,31]. Commonly used biometric features include fingerprints [36], face [37], voice, ECG [38], EKG, vein pattern, iris, palm, DNA, and hand geometry. Because of our focus on health care systems in the current chapter, we will principally focus on the description of ECG-based authentication solutions as a reference.

8.6.2.1 ECG characteristics

ECG signal meets the previously described criteria that should be offered by biometrics; namely, universality, continuity, durability, invulnerability, and uniqueness. In general, ECG signals follow a fundamental morphology, while revealing personalized properties such as beat geometry, the relative timing of the different peaks, and the responses to psychological and physical events [39]. These properties are influenced by age, body build, physiology and geometry of the heart, and gender, making ECG signals distinguishable from one person to another. A typical ECG signal consists of three main waves: a P wave, a QRS complex and a T wave [31]. The P wave illustrates the atrial depolarization, which refers to the electrical signature that causes atrial contraction. The QRS complex is related to the ventricular depolarization. The QRS complex is responsible for contractions of the left and right ventricles. Finally, the ventricular repolarization is represented by the T wave. As illustrated, the most characteristic wave is the QRS complex since it corresponds to the signal with higher amplitudes and is considered as the most recognizable and unique part of the ECG in comparison to the other waves. Therefore, QRS complex is the most frequently used wave during the feature extraction process. A major issue that should be properly addressed is the analysis of the ECG signal and the extraction of significant features from it.

To extract ECG features, the signal must be accurately processed. To this end, several feature extraction solutions have been proposed in the literature including, but not limited to, wavelet transformation [32,40], fuzzy extractor [39], chaotic functions [29], and enhanced fast Fourier transform (FFT) [7]. Due to the durability of ECG signals, the wavelet transformation is not appropriate since the obtained results will not be uniformly distributed and tend to be quasistationary. Therefore, it will be practically impossible to change the generated key incase of leakage and the transformation results cannot be implemented as keys during the data exchange process. Although chaotic functions provide variable results, but such a stability is not yet validated. A fuzzy extractor scheme is proposed in [39] in order to convert noisy nonuniform ECG features into reliably reproducible and uniformly random strings. However, the main limitation of the proposed fuzzy extractor schemes is the lack of reusability and their vulnerability if identical signals are enrolled several times. Processing ECG signals using the FFT method has proven its effectiveness in finding the irregularities in ECG signals [41]. In fact, although the ECG features are highly variable, their detection using FFT can be easily performed with low complexity. The physiological behaviors of the human body will be identified by the peak location index features in the FFT domain, making FFT a good candidate for use in the differentiation of sensor readings collected from one patient to another [7]. In the following, we will describe FFT-based ECG features extraction.

8.6.2.2 ECG features extraction using fast Fourier transform

FFT-based features generation is based on the decomposition of the complex signals to smaller transforms. The decomposed signals are combined to find the resulting transform signal. Low frequencies are removed from the obtained signal using FFT and noises are eliminated after the application of the inverse FFT. The FFT-based key generation can be described as follows:

1. The IMD and the programmer simultaneously measure the ECG signal from the patient, in the frequency domain, for a predefined period of time [16].
2. The WISP RFID tag and the reader apply the enhanced FFT scheme to generate a set of features from the sampled ECG signal as described below:
   a. The measured ECG signal is split into windows.
   b. For each window, a peak detection function and an FFT transformation are implemented in order to return tuples of the form $< k_x^i, k_y^i >$, where k_x^i is the ith peak index and k_y^i is the peak value of the FFT coefficients.
   c. k_x^i and k_y^i are quantized and concatenated to obtain a feature $f^i = [k_x^i | k_y^i]$, which is indexed by the position i.
   d. The programmer and the WISP record a feature vector $F_R = \{f_r^1, f_r^2, ..., f_r^n\}$ and $F_W = \{f_w^1, f_w^2, ..., f_w^n\}$, respectively. An index vector is also recorded to store the positions of the generated features.

At the end of the features generation process, the ECG features computed in both sides (the IMD and the programmer) cannot be entirely equal and differences can be identified because the two signals have been collected at different locations in the patient's body. To this end, to enable the selection of the equivalent biometric features and facilitate the generation of the same key at both sides, a key agreement process during authentication should be performed. In [7] the biometric key generation process is combined with a powerless RFID solution (WISP) to enable to the IMD and the programmer to agree on the same key and to perform a secure access to the IMD either in emergency or ordinary situations [3]. The WISP is implemented as an energy harvesting solution that enforces the secure generation of biometric keys and provides a powerless mutual authentication protocol [12].

8.6.2.3 Powerless mutual authentication using generated biometric keys

Two access modes to implanted/wearable medical devices are generally considered in healthcare: emergency mode and regular mode.

Emergency access mode refers to situations when the health status of the patient is critical and the patient is unable to share his/her credential, the configured credentials are beyond their validity period, or it is the first time a caregiver tries to access the patient equipment. At the programmer side, a new master key is generated and shared with the IMD. To initiate the authentication process, the synchronization request represented by the three-tuple $< N_R, \mathrm{ID}_R, \mathrm{flag} >$ is sent to the WISP from the reader. N_R refers to a nonce aiming to guarantee the freshness of the transaction. ID_R denotes the reader identity and flag is one-bit set to 1. The latter allows indication of whether the access has been acquired using emergency mode. The WISP and the reader in the programmer perform the features generation scheme (previously described) at the same time and generate their correspondent vectors F_W and F_R. Using the generated F_W, a set of random chaff points are generated to hide the genuine features positions corresponding to polynomial values. These chaff points and are expressed by $F'_W = \left\{ \frac{f'^j_W}{f^j_W} \notin F_W, 1 \leq j \leq M \right\}$ and are in the same range of the features computed in F_W [7]. The obtained chaff points are then used to randomly permute the computed features in order to form the vault: $V = \mathrm{RandPermute}(F_W \cup F'_W)$. The vault is then sent to the RFID reader, which will be in charge of identifying the set Q of matching features and their associated indexes I.

Using Q and I the reader implements a one way hash function to compute the secret key K_{bio}: $K_{\mathrm{bio}} = H(Q)$ and to send the message: $< \mathrm{ID}_R, I, \mathrm{HMAC}(K_{\mathrm{bio}}, I|Q|\mathrm{ID}_R) >$ to the WISP. On the IMD side, the same key should be generated. To this end, using I, the set Q should be properly identified by the WISP and the number of the obtained matching features is compared to a given threshold. If it is exceeded, a key $K'_{\mathrm{bio}} = H(Q)$ is computed and matched with the previously computed K_{bio} to check its validity. If the outputs of $\mathrm{HMAC}(K_{\mathrm{bio}}, I|Q|I_{\mathrm{DR}})$ are equal to those provided by $\mathrm{HMAC}\left(K'_{\mathrm{bio}}, I|Q|I_{\mathrm{DR}}\right)$, then K'_{bio} is valid. The WISP generates a nonce N_W and computes the master key: $K = H(K_{\mathrm{bio}}|N_W)$. The first session key $K' = H(K|N_W)$ will be also set up by the WISP and an acknowledgment message $< \{N_W, \mathrm{ID}_W\}_{K_{\mathrm{bio}}}, \mathrm{HMAC}(K_{\mathrm{bio}}, N_R|N_W|\mathrm{ID}_W) >$ will be sent back to the reader. ID_W refers to the WISP identity and $\{-\}_K$ is a symmetric cryptographic function such as AES. Using K_{bio}, the first part of the received message is decrypted by the reader in order to identify ID_W and N_W. A hash-based message authentication code (HMAC) function will be applied over the concatenated value

$N_R|N_W|\mathrm{ID}_W$, in order to check the authenticity of N_W and ID_W. The obtained results from the two parts of the received message are compared with each other. If the results are similar, the reader master and session keys are computed as: $K = H(K_{\mathrm{bio}}|N_w)$ and $K' = H(K|N_W)$. After the achievement of key generation, two messages are sent by the reader. The first message is destined to the WISP and described as follows $<\mathrm{seq}_1, \mathrm{HMAC}(K', N_w|\mathrm{seq}_1)>$. The second message is addressed to the programmer and it contains the session key and the first sequence number $<K', \mathrm{seq}_1>$. At this stage, the WISP is able to verify whether the generated session keys are identical or not. If the same key is generated in both sides, a wake-up request $<K', \mathrm{seq}_1>$ is sent from the WISP to the IMD antenna. The shared master key K is stored in the WISP with the associated information (e.g., the key value, date, and instant generation) and a usage counter is set to 0.

The regular access mode refers to situations when the health status of patient is not critical and therefore the programmer and the IMD are sharing identical credentials (i.e., the master key K) during their validity period. The mutual authentication is initiated by the transmission of the initialization request $<N_R, \mathrm{ID}_R, \mathrm{flag}, \mathrm{HMAC}(\mathrm{ID}_R)>$ from the reader to the WISP. The flag' value is set to 0 (regular mode access). At the reception of the initialization request, the WISP verifies the validity of K and the number of session keys (should not exceed a predefined threshold). If problems related to false key generation, lifetime expiration, or the number of session keys exceeds the threshold, the access is denied. Otherwise, the received message is authenticated by WISP using the stored master key. To initiate data exchange, the WISP proceeds by generation of a nonce N_W, calculation of the session key K, and transmission of the message $<\{\mathrm{Nb}_r, N_w, \mathrm{ID}_W\}_K, \mathrm{HMAC}(K, N_R|N_w|\mathrm{ID}_W)>$ to the RFID reader of the programmer. To obtain Nb_r, N_w, and ID_W, the reader should decrypt the first part of the received message and verify their authenticity using the HMAC function. After authenticity verification, the session key K is computed by the reader using the generated nonce N_W and a message $<\mathrm{Seq}_1, \mathrm{HMAC}(K', N_w|\mathrm{Seq}_1)>$ is transmitted to the WISP. After verifying the freshness of the received message and the keys generated in both sides, the total number of the generated session keys derived from K is incremented by the WISP and as in emergency mode, a wake-up request $<K', \mathrm{seq}_1, \mathrm{Nb}_r>$ is sent to IMD antenna.

8.6.2.4 Establishment of secure communication

Regardless the IMD access mode, a secure communication between the programmer and the IMD can be established using the successfully generated session key K'. In addition, to authenticate the IMD to the programmer while avoiding clogging attacks, an anti-clogging cookie (i.e., an encrypted message including a nonce N_T), and a sequence number seq_1 are attached to each message sent by the IMD. The programmer will decrypt the received message in order to extract the sequence number seq_1 and check its validity by comparing it to the previously sent sequence number. The value of N_T is also extracted and used by the programmer together with the incremented sequence number seq_2 to transmit the encrypted new command in clear text $< N_T, \{cmd\}_{K'}, seq_2 >$.

Upon the reception of this message by the IMD, its freshness should be verified as well as the validity of seq_2. According to [12] three situations are distinguished and three decisions are taken by the IMD:

1. If the received nonce N_T is identical to the one sent previously and the received sequence number seq_2 corresponds to the incremented value of seq_1, then the received message is considered by the IMD as a newly authorized request. After that, the encrypted command will be decrypted, a new nonce N_{T2} will be generated, and the command's response will be computed. The new nonce N_{T2}, the computed response, and the previously incremented sequence number seq_2 are transmitted to the programmer in the form of an encrypted message $< \{N_{T2}, \text{Response}, seq_2\}_{K'} >$.
2. If the received nonce N_T is not identical to the one sent previously and the received sequence number seq_2 does not correspond to the incremented value of seq_1, then the message previously sent by the IMD to the programmer was not appropriately received. Therefore, the previous message will be retransmitted to programmer without exceeding a predefined threshold to avoid DoS attacks.
3. If the received nonce N_T is not identical to the previously sent one but the received sequence number seq_2 corresponds to the incremented value of seq_1, hen the received request is generated by an unauthorized third party (a false data injection attack may have been conducted). In this case, the IMD should discard the received request without proceeding to its decryption.

8.6.3 External/wearable hardware-based solutions

Several works on securing wearable and implanted medical systems have proposed to integrate in the monitoring system an external wearable gadget to act as a gateway node between the IMD and the programmer. The commonly known solutions are cloacker [42], guardian [43], shield [44], and MedMon [45]. The shield plays the role of a jammer-cum receiver that allows using two-antennas to receive and jam synchronously all wireless signals directed to/from the IMD [5,43]. To start the mutual communication between the IMD and the programmer, the programmer should be verified by the shield. If it is considered a genuine party, the shield will initiate the message relay between the IMD and the programmer. This design requires that a secure channel be established between the IMD and the shield Otherwise, the exchanged messages can be easily intercepted by an attacker and jamming attacks could be executed.

The guardian is a gadget worn on the patient's wrist in order to protect the IMD and works differently than the shield. Using asymmetric encryption, it verifies the authorized programmers' authenticity and distributes a session key to the IMD and the programmer using biometric information. When an authentication request is sent from a programmer the IMD, this request is simultaneously received by the guardian. A response to the received request (authentication) will be computed simultaneously by the IMD and the guardian. The IMD initiates a time period to wait for the authentication results that will be communicated by the guardian. If the obtained results are affirmatives, a direct communication between the IMD and the programmer will be established. The guardian will be removed by caregivers in case of emergency to facilitate the access to the medical equipment without affecting the authentication process. The solution proposed by MedMon consists principally on performing anomaly detection by focusing on the physical characteristics of the transmitted signals and reacting by either warning the patient or jamming the signal [46].

8.7 Comparison of security techniques

In this section, we will attempt to analyze the various techniques that are used in WIBSNs for enhancing security and privacy that have been previously described (in Section 8.4) by sketching out their main advantages and limitations. A comparison between these techniques is also provided in Table 8.2.

Proximity-based authentication solutions allow accurate calculation of the distance between two communicating devices using either ultrasonic-bounding, energy-aware authentication, or encryption mechanisms while avoiding distance-shortening wireless attacks (e.g., eavesdropping and spoofing attacks). However, in ultrasonic-bounding access control solutions the radio signals can be easily intercepted by an attacker located within the security range of the device in order to steal the information contained therein. In addition, the implementation of a computationally complex authentication and encryption protocol (e.g., the DH key agreement protocol) consumes high quantities of energy from the device battery. The inability to modify DH parameters after their generation is another drawback. Indeed, when the shared parameters are kept unchanged for a long period of time to derive multiple session keys, they could become vulnerable and therefore, the likelihood of successful cryptanalysis attacks will increase. The wireless communications technologies that have been adopted in IMD design such as Bluetooth, RFID, and MICS are relatively insecure, which makes them vulnerable to attacks such as jamming and man-in-the-middle attacks [7]. The main advantage of using a piezo-element implanted in the patient's body consists in avoiding the implementation of energy-hungry cryptographic solutions that consume the energy of medical devices' battery, yet several weaknesses are also apparent. The connection between caregivers and the medical device is established using a powerless RF platform in the form of a piezoelectric element, which is separately implanted 1 cm under the patient's skin, which may not be cost-effective and is impractical in emergency

Table 8.2 A comparison between the different WIBSN security solutions.

	Mechanisms	Security properties	Attacks targeted	Advantages	Limitations
Proximity-based	Ultrasonic distance bounding	Authentication Confidentiality Integrity Authorization	Eavesdropping Spoofing	Accurate distance calculation	Radio signal interception Complex tasks Energy-hungry tasks Weak authentication Insecure communication technologies
	Energy-aware solutions			Minimal energy overhead	Not-cost effective Impractical in emergency scenarios
Biometric-based approaches	Fingerprinting Retina scan Heart rate Voice recognition	Authentication Confidentiality Integrity Availability	Eavesdropping Jamming	Easy authentication Secure authorized access Redundancy elimination Uniform authentication Minimal energy overhead	Expensive Lack of standardization Possible leakage and high error rates
External/wearable hardware-based solutions	A worn device Clocker Guardian Shield MedMon	Confidentiality Integrity Robustness Authentication	Eavesdropping Node capture Tunneling	Fast authentication, Safety aware solutions Avoid device alteration	Battery consumption Possible key leakage

situations. Similarly to the previous approach, the authentication is not-effective since an adversary who is nearby the patient is able to gain access to the medical devices. In this approach, the security credentials such as the device serial number, its identification number, and secret keys are shared between the IMD and the programmer. However, in case of emergency it might be impossible for the patient to provide these credentials, which makes the access unmanageable. Finally, the two described solutions remain inefficient against resources depletion attacks [12].

Harnessing of biometrics offers several advantages to security and privacy schemes. Indeed, biometrics help to overcome eavesdropping and jamming attacks and facilitate the authentication process inside the resource-constrained WIBSN through the simplification of the patient admission process. Biometrics also help to secure authorized access to medical records, eliminate redundancies, and establish a uniform authentication method. In addition, the overhead caused by traditional key exchange processes can be significantly reduced. Use of biometrics also improves system capability in detection of suspicious events (e.g., misuse of medical data and injection of false data), compromised sensor nodes, and malicious users pretending to be legitimate users. However, biometric technologies may be limited by their affordability (i.e., expensive solutions with lack of standardization) and are affected by several situations including environmental conditions, the patient's daily activity, or accidents, which may affect the patient's body (e.g., a fingerprint-based authentication may not appropriately work if the subject has deformed or mutilated hands after an accident). Other problems that could restrain the success of biometrics are the possible leakage of biometric information and the high error rates due to biometric changes overtime (e.g., retinas and fingerprints) [3].

The third approach consists of using external or wearable hardware-based solutions. The major motivations behind this are: (1) provision of a fail-open access to accomplish the trade-off between accessibility and security. During ordinary access to the IMD, the external hardware will be used to protect the medical system from malicious attacks and illegal activities. During emergency access, unauthorized programmers should be powered off because there is no distributed key or the external hardware should be removed to allow immediate access to the IMD. (2) The mitigation of battery depletion attacks and IMD device alterations. These solutions prioritize the safety of patients and allow fast response times during authentication. However, these attacks can be executed on the external device and despite the fact that its battery can be easily charged and replaced, conducting intensive energy depletion attacks could make it inoperable. Nevertheless, the major limitations of external/wearable hardware-based solutions are associated with security mechanisms endowed on these devices. These mechanism will be effective if the wearable gadget is always worn by the patient, otherwise caregivers with unauthorized programmers or malicious users may be able to successfully access the medical device.

8.8 Emerging security challenges

The spread of integrated/wearable medical sensing technologies spurred development of defensive solutions that try to counter both passive and active eavesdropping attacks while considering the stringent security and design requirements of these systems. The emergence of new technological sensing and health monitoring trends including cloud computing and Internet of Things (IoT) presents several additional security challenges to be considered. Here we describe some of these challenges.

The design of effective lightweight authentication techniques. A practical authentication technique should take into consideration the constrained resources of implanted and wearable sensors while ensuring secure access to their content [47]. In fact, the battery life of some implanted devices (e.g., pacemakers) is expected to last between 5 and 10 years. Therefore, guaranteeing sensor security should preserve their constrained energy without affecting their utility. Several works including [1,9,16,36,45] have addressed this challenge by designing authentication solutions that avoid energy draining programs. These systems either propose energy harvesting solutions using a WISP platform such as in [16] and [1], or integrate an intermediate hardware between the programmer and the medical device such as in [43] and [45]. These two alternatives aim to manage the authentication protocol while using as little energy as possible from the IMD battery. However, in [43] and [45] the security system will be highly dependent on the equipment used. In the case where such equipment is lost by patients or removed by doctors, unauthorized access can be gained. In addition, as the usage of energy harvesting techniques would not always be sufficient against some energy depletion issues, lightweight authentication solutions should be judiciously designed.

The trade-off between speed and security during emergency access to WIBSNs. The designed security systems have considered two types of access that could exist in healthcare applications; namely regular access mode and emergency access mode. During emergencies, the patient's condition is critical and he or she is not able to share credentials [16] or to wear the external gadget [44]. Thus, emergency access solutions prioritize rapid access to the medical device over secure access to its resources. In fact, in external/wearable hardware-based solutions, the emergency access to the IMD resources is accompanied with the elimination of the external equipment. Therefore, unauthorized (accidental or intentional) access to the medical equipment can be achieved. To this end, the trade-off between fast access and secure access to medical devices must be appropriately designed.

Consideration of interference awareness mechanisms. Wireless communications between sensors inside the same WIBSN or between a single WIBSN and the outside world can be characterized by the presence of electromagnetic interferences (e.g., caused by other equipment or jamming attacks). Failure to mitigate these interferences could greatly affect patient safety by hampering the proper sharing of the collected data, delaying the transmission of data with stringent temporal requirements, or increasing data loss rates. The consequences can also have a direct impact on patient's life since they may remain due to nonproperly transmitted data. These interferences cannot be completely avoided since several types of medical and nonmedical equipment could share the same communication channel with WIBSNs. In addition, wearable systems are not only limited to hospitals or residential areas, but can deployed anywhere including in factories, borderlines, airports, and businesses [18]. Therefore, the establishment interference awareness mechanisms and smart access solutions that authorize or deny the acceptance of new systems is required to minimize the risk of attacks and interdevice interference.

The management of massive health data. The massive usage of the newest sensing capabilities and the spread of IoT-based health solutions (e.g., [48–50]) for medical diagnosis and clinical analysis are bringing several opportunities and producing huge volumes of heterogeneous data of various types and with varying constraints. Indeed, beyond traditional sources of health data generated from clinical activities, it is now possible to monitor patients through all kinds of physical devices that are physically or virtually interconnected. Today medical records range from simple medical records collected from wearable or implanted sensors to images and videos transmitted from multimedia sensors and communication equipment (e.g., smartphones). These different types of data are of different value and hence have different security implications for patients. In addition, the large amounts of data increasingly necessitate the design of quick security solutions that enable the detection, identification, and countering of attacks.

The support of data heterogeneity. Collected health data contain a broad range of information (e.g., name, age, address, medications, radiology images, sexual and psychiatric profiles, etc.) that requires special protection (especially of the individually identifiable information) in order to sustain trust between healthcare providers and patients and to protect data ownership confidentiality. Accordingly, by considering issues such as ethical, social, and legal aspects of healthcare systems, the importance of the collected health data, and the risks associated with distributed storage systems offered by the cloud (e.g., private data exposure, data theft or leakage, and data loss), ensuring secure cooperation across domains becomes a challenging issue that should be carefully addressed. Challenges associated to services heterogneity, real-time responsiveness, reliability and availability have to be also addressed.

The design of smart security solutions. Security attacks on healthcare systems are evolving overtime with the emergence of new technological trends. Indeed, more sophisticated and advanced attacks can be conducted by exploiting any vulnerability of the healthcare system. Some of these attacks may not be easily detected with conventional countermeasures schemes. Therefore, more dynamic, smart, and adaptive approaches are necessary. The use of artificial intelligence (AI) and machine learning technologies to provide intelligent security solutions is an active research topic that is addressed in many research reports. Rather than relying on conventional approaches to secure systems, AI-based solutions utilize pattern recognition models that can be built by the systems themselves to automate attack detection and reaction processes using use data from prior similar attacks. On the other hand, to improve sensor utility while optimizing the consumption of their constrained resource, a certain degree of intelligence should be integrated in the configuration programs of the monitoring systems. In fact, the utility of future IMDs will not be only limited to life-sustaining functions.It is also expected that they will allow the collection and storage of accounting data, protection against unauthorized alterations, and detection and toleration of different types of attacks [7]. Consequently, storage and computation optimization solutions must be considered to avoid energy waste by implanted and wearable devices. In addition, intelligence could be related to the design of solutions that prepare for smart digital investigations against attacks targeting WIBSNs. Digital forensics, steganography, and digital watermarking are among the most commonly known approaches that have been discussed in the literature [38].

8.9 Conclusion

WIBSNs are revolutionizing our daily life as a core system in the provision of advanced e-healthcare services. However, these technological developments are generally accompanied with several crucial security and privacy issues. As demonstrated in this chapter, loss of security and privacy properties in the healthcare sector can lead to dramatic and even lethal consequences on patient lives depending on the type of the conducted attack and the nature of the disclosed data. The main objective of this chapter was to give a thorough analysis of the adoption of WIBSNs in the healthcare sector by describing their architecture, identifying possible security and privacy threats, outlining principle security requirements that should be considered, and describing the associated attacks, countermeasures, and related techniques proposed in literature. A set of challenges and future directions related to securing wearable and implanted medical devices were also identified.

References

[1] J. Andreu-Perez, D.R. Leff, H. Ip, G.Z. Yang, From wearable sensors to smart implants-towards pervasive and personalised healthcare, IEEE Trans. Biomed. Eng. 62 (12) (2015) 2750–2762.

[2] Berrahal S. & Boudriga N. (2014), A smart QoS- based traffic management for WBANs, in: Proceedings of the 14th International Symposium on Communications and Information Technologies (ISCIT), Incheon, Korea, pp. 161–165.

[3] S. Berrahal, N. Boudriga, Managing security issues and the hidden dangers of wearable technologies, The Risks of Wearable Technologies to Individuals and Organizations, IGI global, Hershey, PA, 2017, pp. 18–46.

[4] Z. Guanglou, R. Shankaran, M.A. Orgun, K. Saleem, Ideas and challenges for securing wireless implantable medical devices: a review, IEEE Sens. J. 17 (3) (2017) 562–576.

[5] G. Zheng, R. Shankaran, M.A. Orgun, L. Qiao, K. Saleem, Ideas and challenges for securing wireless implantable medical devices: a review, IEEE Sens. J. 17 (3) (2017) 562–576.

[6] Rathore H., Mohamed A., Al-Ali A., Du X. & Guizani M. (2017), A review of security challenges, attacks and resolutions for wireless medical devices, in: In Proceedings of 13th International Wireless Communications and Mobile Computing Conference (IWCMC), Valencia, Spain.

[7] N. Ellouze, M. Allouche, H.B. Ahmed, S. Rekhis, N. Boudriga, Security of implantable medical devices: limits, requirements, and proposals, Secur. Commun. Netw 7 (12) (2014) 2475–2491.

[8] Abouzakhar N.S., Jones A. & Angelopoulou O. (2017), Internet of Things security: a review of risks and threats to healthcare sector, in: Proceedings the IEEE International Conference on Internet of Things (iThings) and IEEE Green Computing and Communications (GreenCom) and IEEE Cyber, Physical and Social Computing (CPSCom) and IEEE Smart Data (SmartData), Exeter, UK.

[9] E. Ghafar-Zadeh, Wireless integrated biosensors for point-of-care diagnostic applications, Sensors 15 (2) (2015) 3236–3261.

[10] N. Dey, A.S. Ashour, F. Shi, S.J. Fong, R.S. Sherratt, Developing residential wireless sensor networks for ECG healthcare monitoring, IEEE Trans. Consum. Electron. 63 (4) (2017) 442–449.

[11] Li C., Raghunathan A., Jha N.K. (2011), Hijacking an insulin pump: security attacks and defenses for a diabetes therapy system, in: Proceedings of the 13th IEEE International Conference on e-Health Networking, Applications and Services (Healthcom), USA.

[12] Ellouze N., Allouche M., Ahmed H.B., Rekhis S. & Boudriga N. (2013), Securing implantable cardiac medical devices: use of radio frequency energy harvesting, In: Proceedings of the 3rd international workshop on Trustworthy embedded devices (TrustED '13), Berlin, Germany, pp. 35–42.

[13] Berrahal S., Boudriga N. & Bagula A.B. (2015), Healthcare systems in rural areas: a cloud-sensor based approach for epidemic diseases management, In: Proceedings of the 7th International Conference on e-Infrastructure and e-Services (AFRICOMM), Benin, pp. 167–177.

[14] S. Yousaf, N. Javaid, U. Qasim, N. Alrajeh, Z.A. Khan, M. Ahmed, Towards reliable and energy-efficient incremental cooperative communication for wireless body area networks, Sensors 16 (3) (2016) 284.

[15] E. Kartsakli, A.S. Lalos, A. Antonopoulos, S. Tennina, M.D. Renzo, L. Alonso, et al., A survey on M2M systems for mHealth: a wireless communications perspective, Sensors 14 (10) (2014) 18009–18052.

[16] Halperin D., Heydt-Benjamin T.S., Ransford B., Clark S.S., Defend B., Morgan W., et al. (2008), Pacemakers and implantable cardiac defibrillators: Software radio attacks and zero-power defenses, in: Proceedings of the IEEE Symposium on Security and Privacy (SP 2008), Oakland, CA, USA.

[17] M. Seyedi, B. Kibret, D.T.H. Lai, M. Faulkner, A Survey on intrabody communications for body area network applications, IEEE Trans. Biomed. Eng. 60 (8) (2013) 2067–2079.

[18] M. Ghamari, B. Janko, R.S. Sherratt, W. Harwin, R. Piechockic, C. Soltanpur, et al., A. Survey on wireless body area networks for ehealthcare systems in residential environments, Sensors 16 (6) (2016) 831.

[19] Eranna U. & Abdul Lateef Haroon P.S. (2017). Signal processing based diagnosis of cardiovascular anomalies, In: Proceedings of the International Conference on Electrical, Electronics, Communication, Computer, and Optimization Techniques (ICEECCOT), Mysuru, India.

[20] H.S.G. Pussewalage, V. Oleshchuk, Privacy preserving mechanisms for enforcing security and privacy requirements in e-health solutions, Int. J. Inform, Manag 36 (6) (2016) 1161–1173.

[21] J. Zhou, Z. Cao, X. Dong, N. Xiong, A.V. Vasilakos, 4S: A secure and privacy-preserving key management scheme for cloud-assisted wireless body area network in m-healthcare social networks, Inf, Sci 314 (2015) 255–276.

[22] C.P. Antonopoulos, N.S. Voros, S. Hey, P. Anastasolpoulou, A. Bideaux, Cyberphysical systems for epilepsy and related brain disorders, Secure and Efficient WSN Communication Infrastructure, Springer International Publishing, 2015, pp. 163–188.

[23] X. Li, J. Niu, S. Kumari, J. Liao, W. Liang, M.K. Khan, A new authentication protocol for healthcare applications using wireless medical sensor networks with user anonymity, Secur. Commun. Netw 9 (2016) 2643–2655.

[24] M. Yaseen, K. Saleem, M.A. Orgun, H. Abbas, J. AlMuhtadi, W. Iqbal, et al., Secure sensors data acquisition and communication protection in eHealthcare: review on state of the art, Telematics Inform. 35 (4) (2017) 702–726. Available from: https://doi.org/10.1016/j.tele.2017.08.005.

[25] Gupta S., Mukheriee T., Venkatasubramanian K. & Taylor T. (2006), Proximity based access control in smart-emergency departments, In: Proceedings of the Fourth Annual IEEE International Conference on Pervasive Computing and Communications Workshops (PerCom Workshops), Pisa, Italy.

[26] Rushanan M., Rubin A.D., Kune D.F. & Swanson C.M. (2014), SoK: security and privacy in implantable medical devices and body area networks, In: Proceedings of the IEEE Symposium on Security and Privacy (SP), USA.

[27] Kim B., Yu J. & Kim H. (2012), In-vivo NFC: remote monitoring of implanted medical devices with improved privacy, In: Proceedings of the 10th ACM Conference on Embedded Network Sensor Systems (SenSys '12), pp. 327–328.

[28] Rasmussen K.B., Castelluccia C., Heydt-Benjamin T.S. & Capkun S. (2009), Proximity-based access control for implantable medical devices, In: Proceedings of the 16th ACM conference on Computer and communications security (CCS '09), Chicago, Illinois, USA, pp. 410–419.

[29] C.K. Chen, C.L. Lin, C.T. Chiang, S.-L. Lin, Personalized information encryption using ECG signals with chaotic functions, Inform. Sci 193 (2012) 125–140.

[30] Singh J.P. & Bilandi N. (2015), Analysis of biometric-based security in wireless body area network (WBAN), In: Proceedings of the International Conference on Information Technology and Computer Science (ITCS), Bali, Indonesia.

[31] Ramli S.N., Ahmad R., Abdollah M.F. & Dutkiewicz E. (2013), A biometric-based security for data authentication in wireless body area network (WBAN), In: Proceedings of the 15th International Conference on Advanced Communication Technology (ICACT), South Korea.

[32] S. Irum, A. Ali, F.A. Khan, H. Abbas, A hybrid security mechanism for intra-WBAN and inter-WBAN communications, Int. J. Distributed Sens. Netw. (2013) 11. Available from: https://doi.org/10.1155/2013/842608.

[33] W. Wang, H. Wang, M. Hempel, D. Peng, H. Sharif, H. Chen, Secure stochastic ECG signals based on Gaussian mixture model for e-healthcare systems, IEEE Syst, J 5 (4) (2011) 564–573.

[34] S. Pirbhulal, H. Zhang, W. Wu, S.C. Mukhopadhyay, Y.-T. Zhang, Heart-beats based biometric random binary sequences generation to secure wireless body, IEEE Trans. Biomed. Eng. 99 (2018) 1.

[35] D.K. Altop, A. Levi, V. Tuzcu, Deriving cryptographic keys from physiological signals, Pervasive Mobile Comput 39 (1) (2017) 65–79.

[36] P. Rajeswari, S. Viswanadha Raju, A.S. Ashour, N. Dey, Multi-fingerprint unimodel-based biometric authentication supporting cloud computing, in: N. Dey, V. Santhi (Eds.), Intelligent Techniques in Signal Processing for Multimedia Security. Studies in Computational Intelligence, vol. 660, Springer, Cham, 2017.

[37] K. Dharavath, F.A. Talukdar, R.H. Laskar, N. Dey, Face recognition under dry and wet face conditions, in: N. Dey, V. Santhi (Eds.), Intelligent Techniques in Signal Processing for Multimedia Security. Studies in Computational Intelligence, vol 660, Springer, Cham, 2017.

[38] S. Nandi, S. Roy, J. Dansana, W. Ben Abdessalem Karaa, R. Ray, S.R. Chowdhury, et al., Cellular automata based encrypted ECG-hash code generation: an application in inter-human authentication system, I.J. Computer Network and Information Security 6 (11) (2014) 1–12.

[39] Huang P., Li B., Guo L., Jin Z. & Chen Y. (2016), A robust and reusable ECG based authentication and data encryption scheme for eHealth systems, In: Proceedings of the IEEE Global Communications Conference (GLOBECOM), USA.

[40] Z. Zidelmal, A. Amirou, M. Adnane, A. Belouchrani, QRS detection based on wavelet coefficients, Comput. Methods Programs Biomed. 107 (3) (2012) 490–496.

[41] K. Akilandeswari, R. Sathya, Feature extraction of ECG signals for early detection of heart arrhythmia, Int. J. Adv. Res. Comput. Commun. Eng. 3 (12) (2014) 8711–8713.

[42] Denning T., Fu K. & Kohno T. (2008), Absence makes the heart grow fonder: new directions for implantable medical device security, In: Proceedings of the 3rd Conference on Hot Topics in Security (HOTSEC'08), USA.

[43] Xu F., Qin Z., Tan C.C., Wang B. & Li Q. (2011), IMDGuard: securing implantable medical devices with the external wearable guardian, In: Proceedings of IEEE INFOCOM, China.

[44] Gollakota S., Hassanieh H., Ransford B., Katabi D. & Fu K. (2011), They can hear your heartbeats: non-invasive security for implantable medical devices, In: Proceedings of the ACM SIGCOMM conference (SIGCOMM'11), Canada, pp. 2–13.

[45] M. Zhang, A. Raghunathan, N.K. Jha, MedMon: securing medical devices through wireless monitoring and anomaly detection, IEEE Trans. Biomed. Circuit. Syst 7 (6) (2013) 871–881.

[46] S. Kulaas, Security belt for wireless implantable medical devices, J. Med. Syst. 41 (11) (2017) 1–9.

[47] T. Hayajneh, B.J. Mohd, M. Imran, G. Almashaqbeh, A.V. Vasilakos, Secure authentication for remote patient monitoring with wireless medical sensor networks, Sensors 16 (4) (2016) 1–53.

[48] G. Elhayatmy, N. Dey, A.S. Ashour, Internet of things based wireless body area network in healthcare. In Internet of Things and Big Data Analytics Toward Next-Generation, Intelligence, Springer, Cham, 2018, pp. 3–20.
[49] N. Dey, A.S. Ashour, C. Bhatt, Internet of things driven connected healthcare, Internet of Things and Big Data Technologies for Next Generation Healthcare, Springer, Cham, 2017, pp. 3–12.
[50] Medical big data and internet of medical things: advances, in: A.E. Hassanien, N. Dey, S. Borra (Eds.), Challenges and Applications, Taylor & Francis, 2019.

Further reading

Michael S.B.R. & Hollick M. (2013), Lightweight energy consumption based intrusion detection system for wireless sensor networks, In: Proceedings of the 28th Annual ACM Symposium on Applied Computing, Portugal.
N. Dey, A.S. Ashour, S. Chakraborty, S. Banerjee, E. Gospodinova, M. Gospodinov, et al., Watermarking in biomedical signal processing, in: N. Dey, V. Santhi (Eds.), Intelligent Techniques in Signal Processing for Multimedia Security. Studies in Computational Intelligence, Springer, Cham, 2017, p. 660.

CHAPTER

9

Smart ambulance traffic management system (SATMS)—a support for wearable and implantable medical devices

Ankur Dumka[1,2] and Anushree Sah[1,2]

[1]*Graphic Era (Deemed to be University), Dehradun, India*

[2]*University of Petroleum and Energy Studies, Dehradun, India*

Chapter Outline

9.1 Introduction

Wearable and implantable sensor technology include technology like wireless body area networks (WBANs), wearable sensors, technologies like Internet of Things (IoT), etc., and have made a tremendous impact in medical fields. While there are a variety of wearable and implantable devices, this chapter focuses on a real-time traffic management system for ambulances and other emergency vehicles which support for effective and fast service for wearable and implantable devices.

Traffic conditions are becoming increasingly worse, and ambulance services are affected because of traffic congestion. The golden hour—the first hour after an accident—is the most critical time for care. This delay can be reduced if with ambulances with smart and intelligent traffic management systems. A variety of traffic

Wearable and Implantable Medical Devices. DOI: https://doi.org/10.1016/B978-0-12-815369-7.00009-4

management systems have been developed for all types of vehicles [1]. Some simple traditional traffic management systems are there that have light emitters that uses infrared and GPS for detecting the presence of ambulance and can also find the real-time density of traffic. Other technologies make use of radio-frequency identification (RFID) tags to identify emergency vehicles and loop inductive methods to determine the number of vehicles. These RFID tags use radio-frequency, which is used to track the position of ambulance. More recent systems used area-based image processing, video-based processing, ultrasonic, infrared emitters, radar, microwaves, etc. to determine traffic in an area. Some systems allow traffic lights to be controlled to allow emergency vehicles to pass intersections. Vehicular ad-hoc networks (VANETs) are considered effective for communication between ambulances and traffic management systems. Fuzzy systems can also be used to monitor real-time traffic at intersections and can also handle the flow of traffic.

The purpose of emergency systems such as ambulances is to provide timely treatment to patients and save them from life-threatening conditions. There are many techniques that can be used to manage traffic conditions so that emergency vehicles such as ambulances, fire trucks, police vehicles, etc., can reach their destinations as quickly as possible [1]. We propose an intelligent and smart ambulance enabled model equipped with IoT sensors, along with GPS technology that can be used for sending the data to specify the location of smart ambulance which is used by smart ambulance traffic management system (SATMS) for providing efficient traffic management. The proposed model (SATMS) is deployed over the cloud to facilitate the traffic condition and other related data to be fetched and analyzed later for taking appropriate decision. The SATMS help in maintaining the traffic plan for ambulance for quick recovery and response which will help for proper implementation of wearable and implantable medical devices.

Management of traffic is a major issue in highly populated metropolitan cities in India like Mumbai, Bangalore, Delhi, etc. The proposed system has the potential to save time and the lives of patients. In this chapter, we discuss traffic management solutions for support of wearable and implantable medical services.

9.2 Case study

While traffic jams and heavy traffic are one of the major problems in big cities, emergency vehicles like ambulances and fire trucks must have a way to get through and reach their destinations in time. The effectiveness of emergency services is based on the time taken by emergency vehicles to reach the location of incident. Delays due to vehicle traffic can lead to loss of life and property. A smart traffic management system that could give priority to emergency vehicles would thus increase the response time and thus efficiency of emergency services.

There has been extensive work done on how efficiently we can use traffic information for determining the sequence of green light, which will give precedence to vehicle which needs emergency response. Nellore et al. [2] and Rajeshwari et al. [3] proposed a traffic management scheme for prioritizing of emergency vehicles. This research focused on giving priority to emergency vehicles. Nellore and Melingi [4] measured traffic conditions using cameras and lane center edges to estimate traffic parameters. Uddin et al. [5] proposed an area-based image processing technique to detect traffic density in an area.

Traditionally strobe emitter– or light emitter–based traffic management systems are used to detect blocked line-of-sight and problem of excessive noise. Farheena and Chandak [6] used RFID technology such as GPS to calculate real-time traffic density in an area. Bharadwaj et al. [7] proposed an inductive loop method in an RFID-based system to count the number of vehicles in a location. Mithun et al. [8] proposed the use of an image-processing-based technique for the detection of traffic patterns in a location. Other researchers have used microwaves, radar, and ultrasound to detect traffic patterns. Abubakr et al. [9] proposed GPS-based technology to detect traffic patterns for efficient traffic management.

Wireless sensor networks (WSNs) technology can be used for providing efficient traffic management, which makes use of embedded sensors which are interconnected with each other for observation and controlling of traffic. Nellore et al. [2] used infrastructure WSNs or vehicular sensor networks (VSNs) for monitoring and management of traffic. He uses video data for management and monitoring of traffic using data collection types such as videos. Abu-Mahfouz and Hancke [10] introduced that uses the location of a vehicle to determine traffic conditions. Fan et al. [11] introduced general packet radio services (GPRS) technology for controlling traffic signals.

Traffic systems like strobe-light systems, acoustic systems, infrared emitters, and radio-based emitter or detector systems have been developed for efficient management of emergency vehicles [12]. These systems make use of preemption which works at the junction where the emergency vehicles are identified by sensors and as the emergency vehicles are identified the proposed technology changes the traffic light signal into green for giving the quick passage to emergency vehicles.

Mittal and Bhandari [13] reported on green wave system for giving priority to emergency vehicles. This system is designed to give green signals to emergency vehicle by activating the green light of the traffic light the emergency vehicle needs to pass. Hegde et al. [14] proposed an RFID- and GPS-enabled system that is used to open the road. Carrying vehicle with emergency response like ambulance. This system uses a mechanism to reduce or shorten the time taken by ambulance to reach the hospital by clearing the lane the ambulance is using. VANET technology was proposed by Jayaraj and Hemanth [15] for effective communication between two different vehicles and for emergency vehicle for traffic control. Chenchela et al. [16] proposed a connection admission control algorithm to improve the performance of traffic management systems. Collotta et al. [17] introduced a fuzzy control mechanism for monitoring real-time traffic and handling congested traffic flow. As noted above, there are numerous techniques for efficient traffic management for ambulances. These approaches helped us develop our emergency vehicle clearance system for vehicles like ambulances to shorten transportation time.

9.3 Proposed design

Our SATMS uses GPS, sensors, IoT, the cloud, etc., for providing to control traffic lights during emergency situations. Increased vehicular traffic in big cities in India and elsewhere is creating congestion at intersection points and causing delays for emergency vehicles or even accidents as ambulances cross red lights at high speed. The proposed model finds the best route based on current traffic conditions. There is a continuous traffic detection and feedback mechanism in SATMS as shown in Fig. 9.1.

9.3.1 Design

The proposed design uses an Information and communications technology (ICT)-enabled ambulance equipped with a GPS-based system and sensors. The GPS-based system is used to provide the coordinates of the position and location of the ambulance to a remote location by making use of sensor-based technology and the cloud. The cloud infrastructure sends information to the SATMS to provide

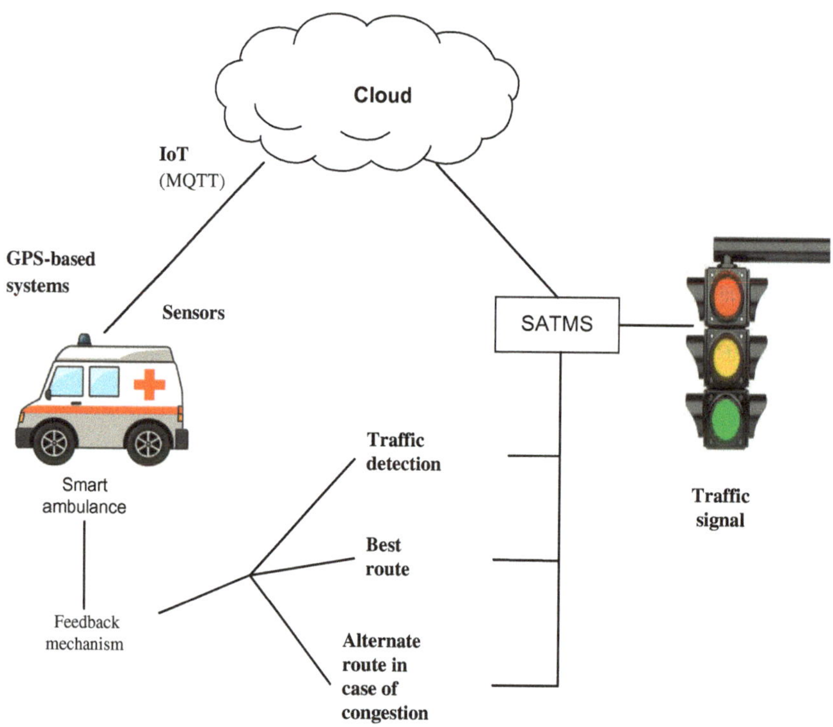

FIGURE 9.1

Proposed design for smart ambulance traffic management system (SATMS).

current traffic conditions with the help of cameras attached at traffic junctions. The SATMS make use of Bellman–Ford algorithm, which works on the concept of distance vector routing protocol for determining the best and shortest route to destination. The SATMS can also be used to provide an alternative route by making use of GIS-based technology. Thus, the three technologies discussed above are used to provide feedback to upcoming vehicle about the traffic conditions and also the best routes to take. The priority-based algorithm proposed and used in SATMS will also help in assigning priorities to the vehicle based on the emergency of vehicles like ambulance, firefighter, police van, normal vehicle, etc. The SATMS is further integrated with traffic light which is used for managing the traffic signals as per the priority of vehicle or emergency of vehicle, thus the vehicle with higher emergency like ambulance will give precedence over other vehicle by opening up the green signal for longer duration of time over the road which is preceded by ambulance.

Our system is dividing into the following phases:

1. Dissemination of data from ambulance
2. Data processing at the cloud
3. SATMS application software

Dissemination of data from the ambulance

IoT-based systems link physical and virtual devices using sensors, actuators, electronic devices, networks, etc. IoT devices is used to collect information from different sensors and process the collected information and send the processed information to the internet IoT is used in:

1. Home automation
2. Smart cities
3. Healthcare systems
4. Smart grids
5. Transportation

In our system, the IoT device has GPS and sensors for getting location coordinates and then sends this data to the SATMS, which is deployed over the cloud as shown in Fig. 9.1.

GPS is a navigation system based on satellites. GPS is a network of satellites that sends details of their position in space back to earth. These satellites are continuously moving in an orbit around the earth. GPS receivers are used to detect the exact position, time, and speed of the device. GPS was introduced in the 1970s and was originally developed by US government for military services during the Cold War. Today it is as common as the mobile phone.

GPS is used to determine the best possible route/routes. In the proposed model, the smart ambulance uses a GPS-based system which is used to locate the position of the ambulance. GPS will send data to RF device and then the data is further sent to the cloud using IoT sensor. The process of sending data from GPS to RF device is shown below [18].

Sender side (GPS to RF device function)

```
if((fGPS=fopen(GPSPort,"r")) == NULL){
        printf("fopen() error could not open %s\n",GPSPort);
                return 1;
        }
        printf("Opening GPS port----- fGPS :%d\n", fGPS);
if((fRF=fopen(RFPort,"w+"))== NULL){
        printf("fopen() error could not open %s\n",RFPort);
        return 1;
        }
        printf("Opening RF port----- fRF :%d\n", fRF);
        flog=fopen(LogFile,"w+");
if((flog=fopen(LogFile,"w+")) == NULL){
        printf("fopen() error could not open %s\n",LogFile);
                return 1;
        }

while(i < lim) {
        bzero(RecvBuf,BSize);
        fgets(RecvBuf,BSize,fGPS);
        n = strlen(RecvBuf);
if (n > 1){
        printf("Msg[%d] Len:%d RF Msg:%s",i,n,RecvBuf);
                        sprintf(SendBuf,"%s:0:%d:0:%s\n",SystemId,i,RecvBuf);
                        printf("SendBuf===>%s",SendBuf);
        fputs(SendBuf,flog);
        fputs(SendBuf,fRF);
        i++;
                }
```

```
Function to save in Log File
while(i < lim) {
        bzero(RecvBuf,BSize);
        fgets(RecvBuf,BSize,fGPS);
        n = strlen(RecvBuf);
        //printf("---o0o---\n",n,RecvBuf);
if (n > 1){
        //printf("Debug:Msg Len:%d RF Msg:%s\n",n,RecvBuf);
        printf("Msg[%d] Len:%d GPS Msg:%s",i,n,RecvBuf);
                        sprintf(SendBuf,"%s:0:%d:0:%s\n",SystemId,i,RecvBuf);
                        printf("SendBuf===>%s",SendBuf);
                        fputs(SendBuf,flog);
                        i++;
                }
```

The data will receive at the receiver as shown in figure below.

Receiver side (functions to receive data)

```
while(i < lim) {
            bzero(RecvBuf,BSize);
            fgets(RecvBuf,BSize,fRF);
            n = strlen(RecvBuf);
            if (n > 1){
                    printf("Msg[%d] Len:%d RF Msg:%s",i,n,RecvBuf);
                    sprintf(SendBuf,"%s:0:%d:0:%s\n",SystemId,i,RecvBuf);
                    printf("SendBuf===>%s",SendBuf);
                    fputs(SendBuf,flog);
                    i++;
            }
```

The following is the log file showing the data transmission from the GPS device to the RF device.

The log file will appear as:

```
admin@localhost:/media/sf_Linux/test/codes
File  Edit  View  Search  Terminal  Help
LC29C:0:0:0:LC29C:0:0:0:$GPRMC,,V,,,,,,,,,,N*53

LC29C:0:1:0:LC29C:0:1:0:$GPVTG,,,,,,,,,N*30

LC29C:0:2:0:LC29C:0:2:0:$GPGGA,,,,,,0,00,99.99,,,,,,*48

LC29C:0:3:0:LC29C:0:3:0:$GPGSA,A,1,,,,,,,,,,,,,99.99,99.99,99.99*30

LC29C:0:4:0:LC29C:0:4:0:$GPGSV,1,1,00*79

LC29C:0:5:0:LC29C:0:5:0:$GPGLL,,,,,,V,N*64

LC29C:0:6:0:LC29C:0:6:0:$GPZDA,,,,,00,00*48

LC29C:0:7:0:LC29C:0:7:0:$GPRMC,,V,,,,,,,,,,N*53

LC29C:0:8:0:LC29C:0:8:0:$GPVTG,,,,,,,,,N*30

LC29C:0:9:0:LC29C:0:9:0:$GPGGA,,,,,,0,00,99.99,,,,,,*48

~
~
~
"RF.log" 20L, 502C
```

9.3.1.1 Sensors

Sensor is used to sense any type of input from the physical environment. This input can be heat, pressure, light, sound, movement, etc. The output of the sensors is usually a signal that can be converted either to human readable form or transferred over a network for further processing or use. The following are some of the types of sensors available today:

1. Image sensors: used in digital cameras
2. Infrared sensors: used to track an object's movement
3. Accelerometer sensors: calculate speed

4. Oxygen sensors: used in vehicle emission control detection systems (finds gasoline/oxygen ratio)
5. Motion sensors: used in security lights, automatic doors, fixtures
6. Photo sensors: detect visible light/infrared/ultraviolet energies
7. Magnetometer sensors: These sensors help google map web service to get your orientation and can be used in compass applications.
8. Barometer sensors: detect pressure
9. Smoke detectors: detect fire or smoke

Sensors are also used with IoT devices to sense what is going on around them. The IoT sensors discussed here are enabled with GPS. These sensors help us find the location and the best route for the smart ambulance to take to the hospital. The direction of movement of the smart ambulance is fetched via a magnetometer sensor that is built into the IoT device.

Data processing in the cloud

In our proposed design we use the SATMS over the cloud making use of a Software as a Service (SaaS) model of cloud computing. The data received from the IoT sensors as shown in Fig. 9.1 is processed by SATMS and stored on the cloud. This stored data can be further processed for analysis of traffic at different places. Cloud technology can be used for effective management of large-scale data.

Geographical information system

Geographical information system (GIS) is a system with the capability of capturing, storing, manipulating, analyzing, and displaying spatial data in an organized form. Thus, GIS is defined as software package which is used for correlating geographical information to attribute retrieved from the data stored in database and vice versa. A GIS can be viewed as a database comprised of geometric elements of geographical space with accurate geometric along with the information needed. Most GIS software has the following components:

1. An input system for collection and processing of data, which is derived from existing maps, remote sensors, etc.
2. A data storage and retrieval system used for organization of spatial data for analysis.
3. A data manipulation and analysis subsystem, which is used for conversion of data into space time optimization and for simulation models.
4. A data reporting system that displays part of a database in the required format.

Mapping is one of the main features of a GIS system. GIS is used to store data and their characteristics on geographical feature. GIS is used for mapping which enables us to map and store cities as points, data of roads as form of lines, boundaries as areas, and scanned maps and aerial photos as raster images. A GIS uses spatial indices to store information used to identify features located in arbitrary positions on the map. GIS mapping can be used for different applications as follows:

1. *Building territories*: Territories and districts can be created using map-based filters and tabular grouping.
2. *Identification of weighted centers*: This is used to find out the center of gravity among points.
3. *Shortest path*: In order to find out the shortest path among all the paths selected, we make use of GIS for finding weather they are ordered or unordered route.
4. *Surface analysis*: This can be used for data querying, surface profiling, view sheds, countering, 3D terrain visualization, Digital elevation model (DEM)/ TIN creation, and calculation of terrain shortest path.

Data classification: It includes several data classification methods for classification purpose as equal weight, equal interval, nested, arithmetic or geometric progression, manual classification by means of range, count or percentage and others. *GPS support*: To read, animate and import data overlay tracks with aerial photos and topographic and vector maps. There are other applications supported by GPS.

Algorithms used for communication

While there are many algorithms that can be used for communication, here we use the Bellman-Ford algorithm, which uses a distance vector to search for the shortest route to a destination. While there are many algorithms that can be used for communication, here we use the Bellman–Ford algorithm, which uses a distance vector approach for finding the shortest route to a destination. It is a versatile algorithm that is capable of handling graphs with negative edges. Bellman–Ford algorithm run in O(|V|. |E|) where V is number of vertices and E is the number of edges. This algorithm calculates the shortest path in a bottom-up manner. First, it finds a path with at-most one edge in the path. Then it considers the shortest path with at most two edges in the path and so on. After the ith iteration of the outer loop, the shortest path with at-most I edges is calculated. The maximum number of edges possible in a simple path will be |V|-1 and hence the outer loop executes |V|-1 times.

We used a priority-based approach for finding the best path based to vehicles like ambulance to give them priority over other vehicle as per need and emergency response needed for any vehicle. This algorithm retrieves information about the priority of vehicle from the proposed system discussed above and maintain vehicle priority as shown in table below, the priority of different vehicle has been set up as follows:

Vehicle	Priority	Back of value
Ambulance	1	Boff 1
Fire truck	2	Boff 2
Police car	3	Boff 3
Normal vehicle	4	Boff 4

We used an average delay time for traffic lights as $T(\text{avg.})$ as A_t, whereas A_m as the maximum time delay for traffic lights to switch from one light to another (green to yellow or yellow to red). For any normal traffic the time slot available with window size of C_w will be $C_w/4$ as considering the case for four corner road. But for priority based vehicle we have to check the back-off time. Which says that for priority 1 the average delay time for the traffic $T(\text{priority1}) < T(\text{avg.})$ which equalizes by a factor K as:

$$T(\text{priority 1}) = T(\text{avg.}) - \left(\frac{K}{S}\right)$$

where variable K is the distance the from ambulance to traffic light and S is the speed of the vehicle. Thus, the time period for the traffic signal turning green for the ambulance will be increased as compared to other vehicles on the other side of signals. For a normal vehicle, the time period is the average time period.

$$T(\text{priority 4}) = T(\text{avg.})$$

This algorithm works on the basis of priority-based scheduling where the lower priority vehicle (which need emergent response) arrives then it will be given priority over other vehicle to move over the road than other vehicle. The flow diagram for this is shown in Fig. 9.2.

Here we had used vehicle as data which need to be transmitted along the road which is used as channel in this algorithm. Based on this, Fig. 9.2 shows that a vehicle wants to send information to traffic signal and will sense the carrier path (road). If the carrier path is busy, then again the request for carrier path will be sent after some time. On the other hand, if the carrier path is available the data (vehicle) will be sent toward the traffic signal and the channel will give a message of channel busy to all other vehicles. All other vehicles have to wait until the data will be send completely by the vehicle using the carrier path (Fig. 9.3).

The proposed system will help in giving priority to vehicle, which need emergency response. The vehicle with high emergency like ambulance will be given lowest priority. The priority of the vehicle will be fetched using IoT-based system to the centralized system and if low priority vehicle will ask for carrier sense (road) for its availability then all other vehicle on the road will have to wait until the low priority vehicle completes its traveling across the road. In the figure, a vehicle ask for carrier sense for its availability. The priority of the new vehicle seeking carrier sense will be checked with the priority of any other vehicle sending data. If the priority of the vehicle seeking carrier sense is less than the existing vehicle sending data or willing to send data the channel will be set free for the low priority vehicle and the information will be passed to the traffic signal to take appropriate action for early passage of a

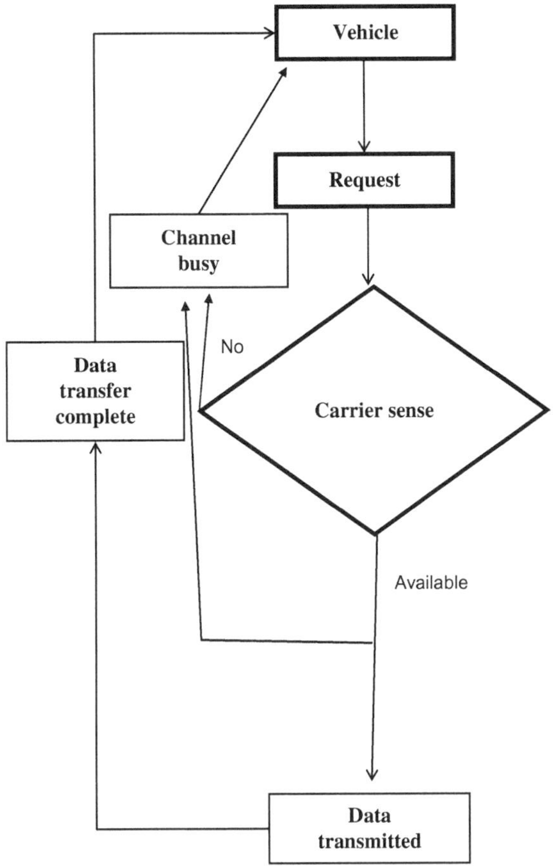

FIGURE 9.2

Normal vehicle data dissemination process using the priority-based algorithm.

high priority vehicle. If the priority of the new vehicle is more than the sending vehicle then there will be channel busy for specified period time till the sending vehicle completes its process. If both vehicles have the same priority the first come first serve process is followed.

Traffic light management

Traffic light management can be done by using inputs from the sensors as discussed previously, we can analyze the traffic movement on the road and accordingly take predictive measurement on the traffic pattern of the road. Based on the predictive analysis of the traffic pattern we can also control the traffic light management using the analytics parameters we receive from the sensor-based technology proposed.

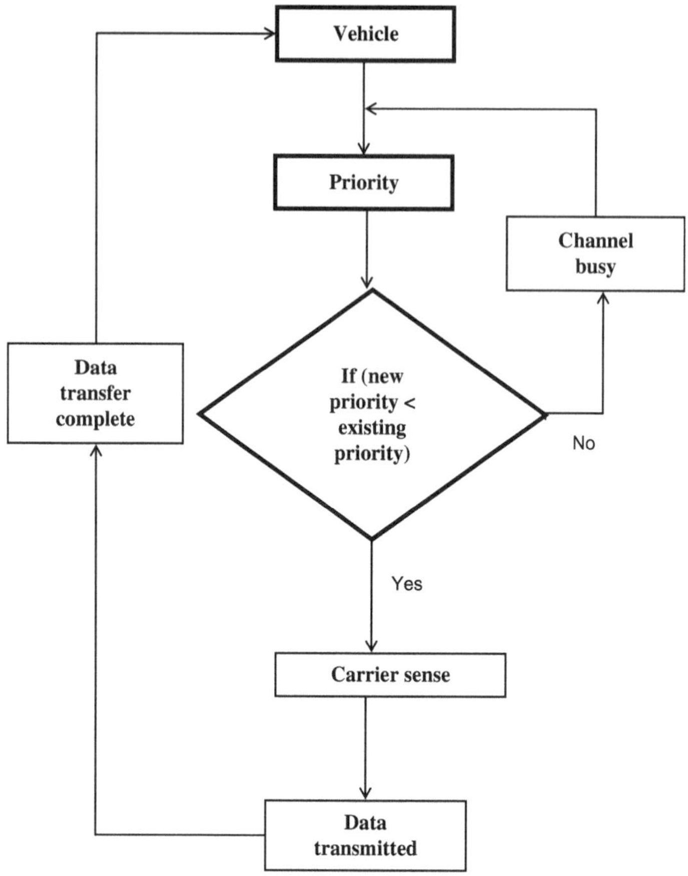

FIGURE 9.3

Data dissemination process for ambulance using the priority-based algorithm.

9.4 Results and discussion

1. Real-time tracking of ambulance: This is one of the advantages of the proposed model—giving the real-time location of an ambulance using GPS to send the information to the cloud where this data is stored for future use. Also, getting the real-time tracking of an ambulance will help doctors and relatives take appropriate actions
2. Traffic detection: Traffic can be detected using the proposed design on a real-time basis using GIS.
3. Best route for ambulance: This is one of the important features of the proposed model where the best route can be detected by means of GIS technology, which provide best possible route for the ambulance to its

destination hospital. Alternate route: This feature can be enabled in the case of traffic jams or any other cause that may delay an ambulance. In these cases, an alternative route will be provided to the ambulance from the source to the hospital to decrease the transportation time.

4. Priority for ambulance using traffic lights: Priority for ambulance using traffic lights: Ambulance is given the priority by using the algorithm proposed which assign lowest priority to vehicle with utmost emergency and based on the emergency of vehicle, the management of traffic light is done to give a safe and quick response passage to high emergency vehicle. Using sensor-based technology we can have prior information for the ambulance arriving at the traffic light and to clear the traffic at that junction for easy and clear passage to ambulance we increase the green signal using real time–based application of IoT for easy passage of ambulance to the required destination.

9.5 Conclusion

This chapter proposed an approach for efficient management of traffic for smart ambulances based on technologies like IoT, cloud computing, GPS, etc., for providing an efficient and real-time solution for traffic management for ambulances and other emergency vehicles.

References

[1] A. Dumka, Smart information technology for universal heathcare, in: Healthcare Data Analytics and Management, Elsevier, 2018, pp. 211–226 (Chapter 8), ISBN: 978-0-12-815368-0.

[2] K. Nellore, V.S. Palepu, M.R.D. Palepu, V.K. Chenchela, Improving the lifespan of wireless sensor network via efficient carrier sensing scheme-CSMA/SDF, Int. J. Eng. Sci. Res. Technol. 5 (2016) 723–732.

[3] S. Rajeshwari, H. Santhoshs, G. Varaprasad, Implementing intelligent traffic control system for congestion control, ambulance clearance and stolen vehicle detection, IEEE Sens. J. 15 (2015) 1109–1113.

[4] K. Nellore, S.B. Melingi, Automatic traffic monitoring system using lane centre edges, IOSR J. Eng 2 (2012) 1–8. 2012.

[5] S.M. Uddin, K.A. Das, A.M. Taleb, Real-time area based traffic density estimation by image processing for traffic signal control: Bangladesh perspective, in: Proceedings of the IEEE International Conference on Electrical Engineering and Information Communication Technology (ICEEICT), Dhaka, Bangladesh, 2015 21–23, pp. 1–5.

[6] S. Farheena, B.M. Chandak, An approach towards traffic management system using density calculation and emergency vehicle alert, IOSR J. Comput. Sci. 4 (2014) 24–27.

[7] R. Bharadwaj, J. Deepak, M. Baranitharam, V.V. Vaidehi, Efficient dynamic traffic control system using wireless sensor networks, in: Proceedings of the IEEE International Conference on Recent Trends in Information Technology (ICRTIT), Chennai, India, 25–27 July 2013, pp. 668–673.

[8] C.N. Mithun, U.N. Rashid, M.M.S. Rahman, Detection and classification of vehicles from video using multiple time-spatial images, IEEE Trans. Intell. Transp. Syst. 13 (2012) 1215–1225. Available from: https://doi.org/10.1109/TITS.2012.2186128.

[9] S.E. Abubakr, O.A. Halla, A.A. Tahani, A GPS based traffic light pre-emption control system for emergency vehicles, in: Proceedings of the IEEE International Conference on Computing, Electrical and Electronics Engineering, Khartoum, Sudan, 26–28 August 2013, pp. 724–729.

[10] A.M. Abu-Mahfouz, G.P Hancke, ns-2 extension to simulate localization systems in wireless sensor networks, in: Proceedings of IEEE Africon 2011, Livingstone, Zambia, 13–15 September 2011, pp. 1–7.

[11] K. Fan, J. Chen, Q. Cao, Design and research on traffic of wireless sensor network based on LabVIEW, in: Proceedings of the 2nd International Symposium on Computer, Communication, Control and Automation (3CA 2013), Singapore, 1–2 December 2013, pp. 6–9.

[12] N. AI-Ostath, F. Selityn, Z. AI-Roudhan, M. EI-Abd, Implementation of an emergency vehicle to traffic lights communication system, in: Proceedings of the 7th International Conference on New Technologies, Mobility and Security (NTMS), Paris, France, 27–29 July 2015, pp. 1–5.

[13] K.A. Mittal, D. Bhandari, A novel approach to implement green wave system and detection of stolen Vehicles, in: Proceedings of the 2013 IEEE 3rd International Advance Computing Conference (IACC), Ghaziabad, India, 22–23 February 2013, pp. 1055–1059.

[14] R. Hegde, R.R. Sail, S.M. Indira, RFID and GPS based automatic lane clearance system for ambulance, Int. J. Adv. Elect. Electron. Eng. 2 (2013) 102–107.

[15] V. Jayaraj, C. Hemanth, Emergency vehicle signaling using VANETS, in: Proceedings of the IEEE 17th International Conference on High Performance Computing and Communications (HPCC), 2015 IEEE 7th International Symposium on Cyberspace safety and Security (CSS), and 2015 IEEE 12th International Conference on Embedded Software and Systems (ICESS), New York, NY, USA, 24–26 August 2015, pp. 734–739.

[16] V.K. Chenchela, V.S. Palepu, K. Nellore, M.R.D. Palepu, Improving the quality of service (QOS) of connection admission control mechanism (CAC) using two dimensional queuing model, IOSR J. Electr. Electron. Eng. 11 (2016) 17–27.

[17] M. Collotta, L.L. Bello, G. Pau, A novel approach for dynamic traffic lights management based on wireless sensor networks and multiple fuzzy logic controllers, Expert Syst. Appl. 42 (2015) 5403–5415. Available from: https://doi.org/10.1016/j.eswa.2015.02.011.

[18] A. Dumka, Smart metering as a service using Hadoop (SMAASH), in: Computational Intelligence Application in Business Intelligence and Big Data Analytics, CRC Press, Taylor & Francis Publishers (Chapter 8), 2017, ISBN 9781498761017.

Further reading

A. Dumka, A. Sah, Smart ambulance system using concept of big data and Internet of Things, in: Healthcare Data Analytics and Management, Elsevier, 2018, pp. 155–176 (Chapter 6), ISBN: 978-0-12-815368-0.

K. Nellore, G.P. Hancke, A survey on urban traffic management system using wireless sensor networks, Sensors. 16 (2016) 157. Available from: https://doi.org/10.3390/s16020157.

CHAPTER

10 Internet of things-linked wearable devices for managing food safety in the healthcare sector

Farah Bader[1] **and Sandeep Jagtap**[1,2]

[1]*Centre for Sustainable Manufacturing and Recycling Technologies (SMART), The Wolfson School of Mechanical, Electrical and Manufacturing Engineering, Loughborough University, Leicestershire, United Kingdom*

[2]*The National Centre for Food Manufacturing, University of Lincoln, Holbeach, United Kingdom*

Chapter Outline

Wearable and Implantable Medical Devices. DOI: https://doi.org/10.1016/B978-0-12-815369-7.00010-0

10.1 Introduction

The Press Association, in the reports obtained in 2016 from the Food Standards Agency, exposed poor food standards at 400 hospitals, hospices, care homes, nurseries, and school clubs in the United Kingdom. The report stated that these institutions need significant improvements in their food preparation practices. Much of the food was prepared in kitchens infested with pests, molds, unclean surfaces, and unsanitary utensils, all of which were highly unacceptable, per ITV News [1]. These institutions cater to the most vulnerable people in society, and the standards of food hygiene should be of paramount importance. The systems currently in place used to monitor food hygiene are mostly paper-based, which are noneffective, nontransparent, and sometimes manipulated by the preparation staff [2]. Therefore, to offer a system that is transparent and can ensure that food has undergone stringent quality checks is a relevant need for both academic and National Health Service (NHS) management teams [3].

The Internet of Things (IoT) concept, which has been around for a few years, can help in addressing the food safety issue. IoT can ensure transparency [4] throughout the healthcare's food supply chain by improving efficiency and alerting the stakeholders if there is a sudden deviation from set standards. Every single movement of food can be monitored, and hazards arising from the mishandling of food can be eliminated. Some stages of the food supply chain such as production, warehousing and transportation have already adopted IoT [5]. In production, Radio-Frequency Identification (RFID) technology is used for scanning products, providing further guidance in the assembly process [6] and capturing real-time information of the product and processes [7]. Sensor technologies are used to monitor and record temperatures, microbiological information, and food quality parameters [8,9]. Other applications of IoT include real-time production scheduling [10], real-time wireless manufacturing [6], material handling [11], and operational and resource optimization methods [12].

Many architectures and IoT-based applications have been proposed for managing food safety in the food supply chain, yet some challenges remain. Despite difficulties, an attempt has been made to adopt a suitable IoT-based architecture for maintaining food safety in healthcare. To implement the IoT architecture, it is of the utmost importance to understand the food preparation system used by healthcare institutions, the benefits of adopting IoT, and the areas within the preparation process that are eligible to best embrace wearable IoT devices to improve the safety of food processing.

This chapter begins by exploring the basics of food hygiene and how to achieve food safety to avoid possible life-threatening foodborne diseases. The current methods of food hygiene monitoring and hazard assessment are also explored before highlighting the significance of improving these methods within healthcare food service facilities. The final part of this chapter discusses the proposed system design.

10.2 Background and context

To fully understand the hazards that can be found in food products, we must first understand the basics of food deterioration. The following section explores the origins of food hygiene science, and how failing to sustain food safety can lead to detrimental health effects.

10.2.1 Food hygiene and safety

Food hygiene includes more than just cleanliness; it also indicates the proper use of policies, practices, and procedures to protect food from contamination, prevent the multiplication of bacteria capable of causing food poisoning/spoilage, and ensure the destruction of disease-producing microorganisms by thorough cooking [13]. Ensuring that a food product is safe to eat by controlling the hygiene during all stages of food processing and preparation is vital to ensuring food safety. The absence of such food hygiene will endanger the food's safety and expose consumers to foodborne illnesses, which can be fatal [14].

Food safety entails the production of food products that are fit for human consumption and are safe to eat, meaning they are free of any disease-causing agents, toxins or foreign objects. The aim is to produce such food while maintaining the quality needed to ensure it is desirable to eat, having the desired taste, color, and texture.

10.2.2 Food spoilage and deterioration

Food deterioration is the decline of a food product's quality, texture, flavor, odor, or color, causing the food to both appear as and become an unpleasant object. Food spoilage leads to food deterioration and spoilage is attributed to either natural decay or microorganism contamination. Spoilage may occur naturally with moisture loss, enzyme action, or chemical changes within the food and/or the interactions with its environment. If the cause of deterioration is spoilage by enzymes, then the food will experience changes in texture and flavor. If food spoilage occurs via microorganism contamination, foods will become infected with bacteria, molds, or yeasts. If the food develops unusual coatings on its surface, this is attributed to yeast hazes, molds, or slime layers, and these coatings are referred to as biomasses. Presence of microorganisms may also cause curdling, souring, or blowing (e.g., inflated cans), such spoilage is referred to as "major product spoilage." Table 10.1 lists the two main factors of food spoilage: spoilage types and spoilage effects on food that cause food deterioration.

10.2.3 Foodborne disease

Any illness resulting from the consumption of contaminated food products will result in foodborne diseases. Several events must occur before a foodborne

Table 10.1 Food spoilage factors, types, and subsequent food deterioration.

Food spoilage factor	Type and subsequent food deterioration
Natural decay	1. Moisture loss: Water leaves food products when exposed to air. Causes loss of appearance and texture. 2. Enzymatic action: When food products are processed, some enzymes are released, which chemically react to the air they are exposed to (i.e., enzymatic browning). Brown pigmentation begins to develop, which causes discoloration of food products. 3. Chemical changes: Extrinsic factors such as oxygen, sunlight, and temperature trigger chemical changes in food products during processing. Flavors and colors are severely affected, causing offensive odors and tastes, as well as unpleasant appearance.
Microorganism contamination	1. *Bacteria*: Found in air, water, plants, animals, and people. No changes to food products, but consumption of pathogenic bacteria-contaminated food will cause food poisoning. 2. *Molds:* A type of fungi that absorbs nutrients from its environment, allowing it to grow on the surface of foods and produce spores, which include harmful toxins. Changes the appearance, texture, and color of food products; consumption of molds may cause poisoning. 3. *Yeasts*: Single-celled fungi, yeasts grow in warm and moist environments, typically in food products with high sugar content. Their growth is associated with the production of gases, which can lead to bloated cans or popping of bottles.

disease can develop. First, the food must encounter a pathogen to become contaminated. Second, for the pathogen to grow and lead to disease, the food must be subjected to temperature abuse for a prolonged period of time. Finally, to result in illness and disease symptoms, the consumer must have ingested a high level of the hazardous substance. Foodborne illnesses may result in poisoning from infection, intoxication, or allergic reactions. Symptoms of food poisoning include nausea, diarrhea, abdominal pain, weakness, fever, bodily aches, and chills [15]. In the United Kingdom, *salmonella*, *Escherichia coli* and viruses are the biggest causes of food poisoning, all of which are associated with raw meats, poultry, and shellfish, as well as unpasteurized milk and ready-to-eat meals [16]. An in-depth analysis of each type of poisoning follows.

1. *Infection*: infections can be caused by a variety of bacteria, viruses, and parasites. For the pathogen to grow, the contaminated foods must be subjected to temperature abuse for an extended time. Bacterial infections are caused by pathogenic cells that thrive in the digestive tract after surviving the acidity of the stomach. Toxin-mediated infections occur in consumers who ingest foods carrying fully grown microbes that settle in the intestinal tract, where they continue to grow and produce toxins; in essence, the food carries live pathogens, which colonized within the body. Some infections are caused by consuming foods that already contain a toxin, which once ingested cause life-threatening illness.
2. *Intoxication*: a gravely serious illness caused by the consumption of foods that are contaminated with extremely high levels of pathogens, or that have been contaminated for an extended period of time. The high level of pathogens or the elongated contamination time allows for the production of toxins. These pathogens settle and multiply rapidly in the intestinal wall lining and grow to produce more toxins.
3. *Allergens*: the unintentional presence of one of the 14 listed food allergens in food products can cause allergen poisoning in consumers who are have these allergies. The Allergens are celery cereals containing gluten, crustaceans, eggs, fish, lupin, milk, molluscs, mustard, peanuts, sesame seeds, soybeans, as well as sulphur dioxide and sulphites, and tree nuts. These allergens may be present due to cross-contamination, and their consumption may cause anaphylactic shock wherein the throat swells, causing difficult breathing and possible loss of consciousness. Other symptoms include sneezing, itching, watering eyes, chest tightness, wheezing, rashes, and swelling of face, lips, eyes, and tongue [17].

There are several factors that contribute to the development of foodborne illnesses in food products. The risk factors include but are not limited to: improper cooking, cooling, storage, or reheating conditions/temperatures [18], long storage times between preparation and serving ($>$12 hours), unhygienic or infected food handlers, consumption of food products that are prone to contamination when raw and cross-contamination, and the use of leftover preprepared food products. Cross-contamination in food service facilities is largely attributed to the food handler's poor personal hygiene [16]. Table 10.2 lists the risk factors and their roles in facilitating foodborne illnesses as adapted from Ref. [16].

10.2.4 Food safety hazards

Although the consumption of spoiled foods leads to foodborne illness that may be life-threatening, it is only one part of several other factors that are taken into consideration when assessing food safety hazards. The hazards in food differ based on type, origin, recipes, and preparation procedures. To fully

Table 10.2 Risk factors that may lead to a foodborne illness.

Risk factor	Foodborne illness facilitated by
Improper cooking	Survival
Improper cooling	Growth
Improper storage (under heat)	Growth
Improper reheating	Growth and survival
Long storage time between preparation and serving (> 12 h)	Growth
Infected food handlers	Contamination
Contaminated raw food	Contamination
Cross-contamination	Contamination
Leftover preprepared food products	Contamination, growth, and survival

Source: *Adapted from B. Ray, A. Bhunia, Fundamental Food Microbiology, fourth ed., CRC Press, 2007 [16].*

accumulate all food hazards to include both intrinsic and extrinsic factors, it is best to display them under three categories: chemical, physical, and microbiological hazards [19].

10.2.4.1 Chemical

Chemical hazards in food include toxic substances that may be produced naturally or those that are added intentionally or unintentionally to food products [19].

1. *Naturally occurring*: living organisms such as bacteria and molds in food produce substances that may be toxic; some of these may only result after processing. These toxins can cause illness and may lead to death, therefore, they must be monitored and regulated.
2. *Intentionally added*: substances that may be added for a practical use in the manufacturing, processing, and preservation of food products. These are not considered food itself or used as a food ingredient in any way. An example is nitrates in meat: these substances are regulated by legislation, and they cause illness when the levels added to meat exceed the safety limit.
3. *Unintentionally added*: substances used in the environment surrounding the food such as cleaning agents, which contaminate the food during processing.

10.2.4.2 Physical

Physical hazards in food can be defined as the presence of foreign objects such as wood or glass due to poor design and/or control of the production and preparation environment. These hazards include the accidental addition of items from the food's immediate environment by food handlers or servers, as well as items that may contaminate the food.

The majority of these items are small and go undiscovered by the human eye. Some food service facilities will have detection machinery while others will not detect the presence of the hazard until after service [19].

10.2.4.3 Microbiological

Microbiological hazards are identified by type, name, foods associated with it, and the threat it poses to groups of individuals. As mentioned above, microorganisms are found all around us – they are present in the air, water, and naturally on most food products. These microorganisms include yeasts, molds and bacteria (as seen in Food Spoilage section above), as well as viruses. Table 10.3 lists some of the pathogens most prominent in microbiological outbreaks.

The ideal temperature for microorganisms growth is 37°C; however, growth is still possible at other temperatures, as seen in Fig. 10.1. The time it takes for microorganisms to multiply differs, however, in order for pathogenic microorganisms to survive, the food must be subjected to temperature abuse, inadequate cooking or incidental contamination in some way. To ensure microbiological safety, food products must be kept at safe moisture levels. Sanitation and cleaning are highly important, and should be done in ways that do not compromise the quality or safety of the food products.

10.2.5 Management of food hazards

Prevention of foodborne illnesses is possible by using food from safe sources along with proper storage conditions, temperature regulation, and time controls. Succeeding at the prevention of foodborne illnesses is facilitated by imposing strict employee personal hygiene policies, as well as policies for the cleaning and sanitation of the work environment. These factors are considered basic to any food production facility, ensuring the use of safe water and clean equipment (Ray and Bhunia, 2008).

Such proper conditions are only successfully executed when employees are trained to follow the appropriate measures. Because of the challenges affecting the safety of food products, there are common methods of maintaining good hygiene as well as assessing the probable hazards in order to avoid them. The two general methods used to manage any food production facility are Good Manufacturing Practice (GMP) and Hazard analysis and critical control point (HACCP).

10.2.5.1 Good Manufacturing Practices

GMPs include the general conditions of manufacturing, packaging, and storage needed to maintain hygiene of food for human consumption. At the food level, GMPs are maintained by guaranteeing that the raw food material selected is clean and suitable for human consumption. Any foods that are known to contain natural, unavoidable biological material that may be harmful to consumers must be checked to ensure levels of unavoidable material comply with set standards.

Table 10.3 Microbiological hazards in food [20,21].

Microbiological hazard	Type
Bacteria	1. *Bacillus cereus* Found in cooked meat, poultry, rice, potatoes, pudding, soups, and vegetables. Cause mild food poisoning but pose a severe hazard for vulnerable patients. 2. *Clostridium botulinum* Found in unprocessed canned foods, minimally heated chilled foods under temperature abuse, raw or cooked herbs or vegetables immersed in oil, and honey. Their spores survive cooking and thrive in improper storage. Pose as severe poisoning hazard to vulnerable patients and the general public. 3. *Escherichia coli* Found in undercooked meat as well as in cooked meat (via cross-contamination from raw meat), unpasteurized milk and milk products, unpasteurized fruit juices, raw vegetables, and ready-to-eat foods. Once grown, they produce the shiga toxin, which poses a severe hazard to vulnerable patients and the general public. 4. *Staphylococcus aureus* Found in food that is contaminated and subjected to time/temperature abuse, ready-to-eat foods touched with bare hands and refrigerated inadequately. Causes intoxication and poses a hazard to the general public. 5. *Listeria* Found in refrigerated ready-to-eat food products, especially those containing cheese or meats. Spores are killed under cooking temperatures but may survive and slowly grow at refrigeration temperatures. Causes food poisoning and poses a severe threat to infants, young children, elderly, and vulnerable patients. 6. *Salmonella* Found in poultry, eggs (cooked, undercooked, or raw), and egg products. Causes food poisoning with a severe threat to infants, young children, elderly, and vulnerable people. 7. *Campylobacter* Found in dairy, poultry, red meats, and all of their products. Spores are killed by thorough cooking. Causes infection and diarrheal illness. Mild threat in general public but may be fatal in infants, young children, elderly, and vulnerable people.
Viruses	1. *Norovirus* Found in shellfish, fresh fruit and salad vegetables, and ready-to-eat foods contaminated by handlers. Causes infection in consumers and poses a threat to vulnerable people and the general public. 2. *Hepatitis A or E Virus* Found in shellfish, fresh fruit and salad vegetables, ready-to-eat foods contaminated by food handlers and pig meat. Spread mainly by food or water contaminated by the stool of an infected person; causes infection and poses a severe threat to vulnerable patients and the general public.

(*Continued*)

Table 10.3 Microbiological hazards in food [20,21]. *Continued*

Microbiological hazard	Type
Protozoa	1. *Cryptosporidium species* Found in raw fruit and vegetable contaminated by irrigation water, unpasteurized milk and juice, shellfish, ready-to-eat foods contaminated by the handler. Leads to food poisoning and poses a severe hazard for vulnerable patients. 2. *Toxoplasma gondii* Found in raw and undercooked meat (pork, lamb, venison), unwashed fruit and vegetables, or food contamination from hands that have touched cat litter. Causes food poisoning that may be a severe hazard for vulnerable people.

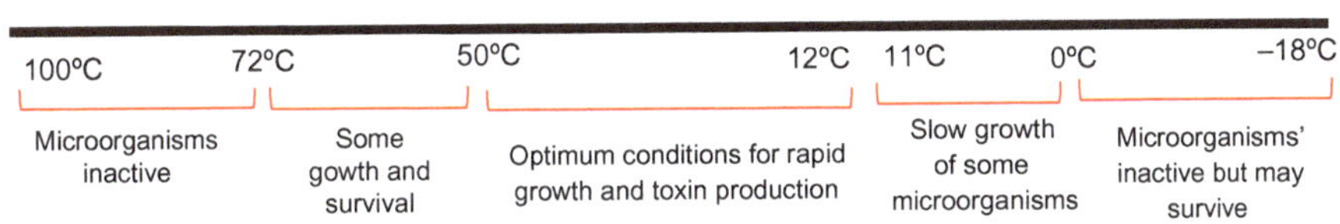

FIGURE 10.1

Status of microorganisms at different temperatures.

Monitoring is also important to ensure levels of these materials are not affected by production and processing operations involved in their preparation [22]. At the manufacturing and equipment level, GMPs are implemented through ensuring the upkeep of employee, facility, and equipment hygiene. This includes taking into account equipment design, construction, and maintenance, which is essential for hygienic food, free from microbiological, chemical, and physical hazards. Other precautionary GMPs may include designing equipment for easy cleaning (e.g., smooth surfaces, no grooves or crevices where contaminants can accumulate), the use of assays to ensure microbial cleanliness and the removal of airborne contaminants via air filtering systems [23].

10.2.5.2 Hazard Analysis and Critical Control Points

If a production facility recognizes all of the possible food hazards and the nature of how they occur, they will be able to predict the probability of the hazards manifesting into outbreaks, and either prevent or control their occurrence. With this aim, HACCP is a management tool that provides a structural, systematic approach to the identification of hazards and the methods of control, beyond what is achievable by traditional testing and inspection approaches of quality control. This allows for a preventative approach to ensure food quality is deemed safe for human consumption. This tool allows for comprehensive identification, evaluation, and control of all hazards significant for food safety. There are four main stages of HACCP: planning and preparing, plan development, implementation, and finally maintaining the HACCP [19]. Within these steps, HACCP uses the following principles:

1. Conduct hazard analysis
2. Determine Critical Control Points (CCPs)
3. Determine critical limits
4. Create a monitoring system for CCPs
5. Create corrective action for when CCP control fails
6. Create a procedure to verify and confirm effective HACCP implementation
7. Ensure documentation and recording of all procedures

CCPs define a step at which a food safety hazard may occur, meaning control and monitoring is essential at these points to ensure immediate action is taken when required. The control measures defined at CCPs must eliminate a food safety hazard or reduce it to an acceptable level. At least one step in any food processing or preparation facility must be a CCP to ascertain that the probability of a potential hazard occurring stays as low as possible. Each CCP identified must be accompanied by corrective actions to be taken when required; failure to control CCPs may lead to harming consumers. This approach applies to all hazards, using systematic and prevention approaches to hazards allow food preparation institutions to enhance control and integrate quality assurance in their production. This allows for better compliance with the food industry standards and legislation, as well as the production of safe foods without reliance on end-product testing.

Planning for HACCP requires sufficient time, proper employee training, and commitment as it includes a lot of data collection and use of resources [22]. Many institutions may require external assistance for this degree of planning due to limited knowledge. Several HACCP plans fail in implementation due to ineffective employees who do not to possess the appropriate skills required [24].

However, the highest levels of food safety are only possible when HACCP is applied along with all prerequisite programs (PRPs). PRPs include all the general policies, practices and processes needed for safe foods in constant operation, which includes GMP in order to ensure compliance with ambitious food hygiene standards. PRPs for microbiological hazards will generally focus on monitoring temperatures throughout the preparation, cooking, and storage of food. Chemical and physical hazards may require more visual monitoring with various CCPs [23].

10.3 Healthcare food service facilities

Fig. 10.2 demonstrates the movement of food typically found in NHS hospitals. It starts with the receipt of the food from the suppliers in the form of ingredients until it is processed and served to patients or visitors within the healthcare environment. NHS hospitals receive food ingredients from authorized suppliers who are high-quality raw material suppliers. The food received is checked for quality, quantity, price and other conformance such as temperature, microbial reports, etc. Then, the food is transferred to storages where it is appropriately stored in chillers, freezers or ambient area. Depending on the preorders, a quantity of raw materials in the storage are picked and decanted in preparation for the next day. The rule "first-in-first-out" (FIFO; that is, oldest stock is to be used before newer stock) is followed, which ensures stock rotation. Frozen foods (meats, fish, etc.)

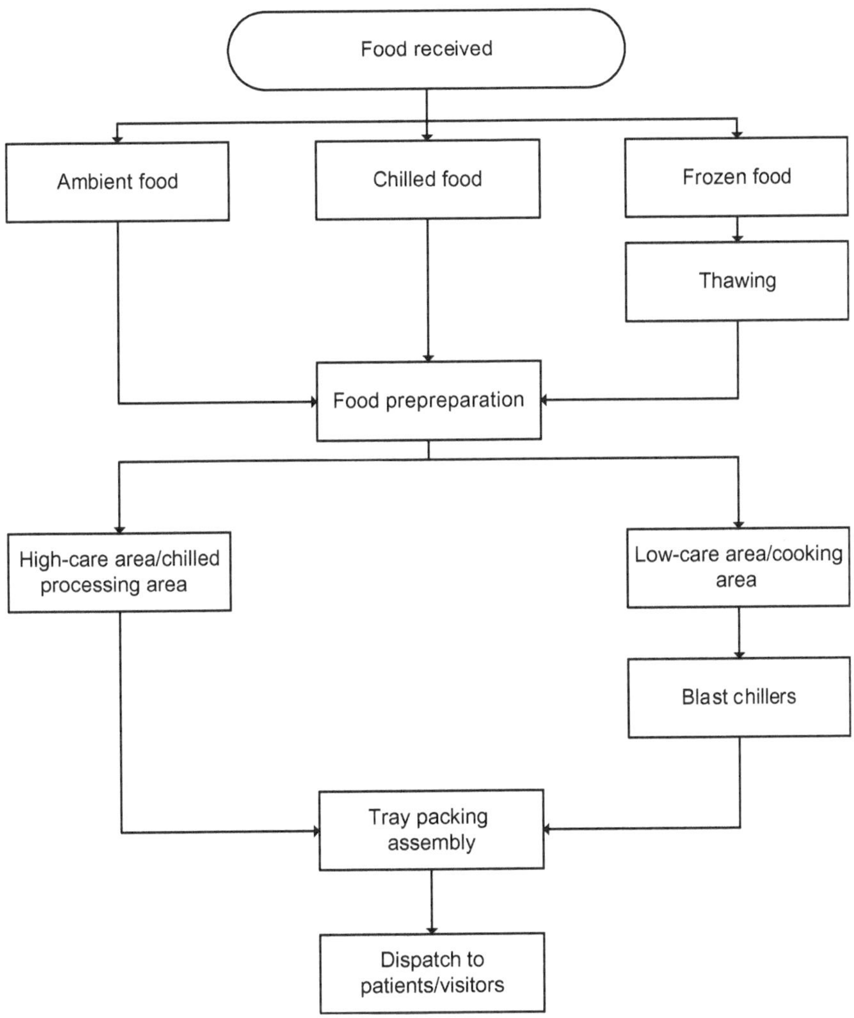

FIGURE 10.2

NHS food preparation system.

are thawed before they are sent to the preparation department. In the preparation area activities such as washing, dicing, cutting, chopping, and marinating of food is carried out. The prepared food is sent to either low-care or high-care areas for further processing. The food which needs cooking is transferred to low-care area whereas food that does not need any processing (salads, sandwiches, etc.) and is suitable for human consumption is sent to the high-care area. The cooked food is cooled down rapidly using blast chillers before they can be packed along with high-care food in the tray pack area. From there, the food trays are dispatched to various wards and canteen facilities.

Overall, the handling of food during preparation and storage within food service facilities exposes food to most of the contamination it will encounter. It is important to remember that hospitals treat many highly immune-compromised patients, including organ-transplant and cancer patients, the elderly, and infants [13]. Thus life-threatening outbreaks can cause many avoidable deaths. Staff and patient visitors may also fall ill as a result, which compromises the quality of the services and support provided to the patients. Although HACCPs should result in a straightforward preparation system, several challenges have influenced food safety in hospitals [21]. The following is a list of factors outlined as requirements for achieving high levels of food hygiene and safety in healthcare settings.

1. *Trained staff*: employees in the food service facilities that are not properly trained and do not receive enough support from managers to monitor food safety will lead to outbreaks. Food handlers must be trained appropriately, and it is imperative that they maintain personal hygiene.
2. *Storage and distribution control*: meals are prepared and stored at the same time but may be distributed at different times, meaning they all have different holding times and may suffer temperature abuse that increases microbial hazard risk. It is important to serve foods within 15 minutes of regeneration (i.e., heating).
3. *Control of food*: ensure food is clean and from a suitable source. Cook all foods to their respective optimal internal temperature and hold this temperature until food is cooked thoroughly, inactivating any pathogens that may exist.
4. *Facility design and control*: equipment must be easy to clean and sanitize. The facility should allow for separation of processing steps so that cross-contamination between raw foods and ready-to-eat foods can be easily avoided. Separate equipment and utensils should be used for each food item, and these should be properly cleaned and sanitized.

10.4 System design

10.4.1 IoT-based architecture for food safety in healthcare

Based on the systems and IoT applications in food supply chains and other fields, the possibilities for managing food safety in healthcare are proposed as shown in

Fig. 10.3. The proposed system consists of four layers: the sensing layer, the network layer, the service layer, and the application layer.

- The *sensing layer* consists of physical objects and corresponding smart devices becoming "smart" by collecting and exchanging the data. From the food safety viewpoint, devices can collect data such as temperature, location and quality, as well as track every staff movement through various sensors, cameras, RFID tags, Quick Response (QR) codes, or barcodes. These devices are to be worn by staff as a part of their uniform or attached to the equipment such as food trolleys, protective glasses and other heavily used supplies. Sensors help to identify dangerous bacteria such as *salmonella* and *E. coli*, as well as to detect food spoilage from the gas emitted by the food products. Sensor-based networks form an integral part of this layer.
- The *network layer* is a medium between sensing and service layer. It is responsible for connecting the smart objects to servers and transmitting sensor

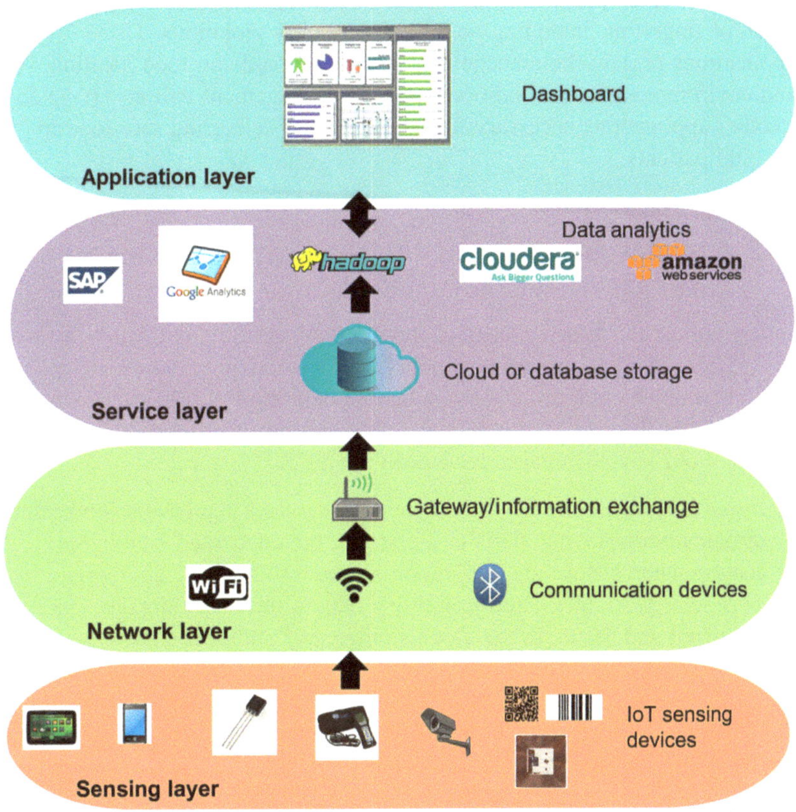

FIGURE 10.3

IoT architecture.

data to the service layer for data processing and transferring purposes. It uses WiFi, Bluetooth, and Ethernet technology to execute the collection and transfer of data. It consists of two basic components – gateway things and gateway networks, where gateway things consist of processor units and gateway networks consist of devices such as protocol translators, rate converters, and signal translators. This layer is responsible for securing the network, giving authorization, controlling access, and encryption infrastructure related to food safety.

- The *service layer* performs two functions: first, storing data on a local database or cloud (digital memory storage). Secondly, it performs data analytics on the collected data to extract meaningful information with regards to food safety. It uses common analytical platforms such as SAP, Cloudera, Hadoop, Google analytics, etc. These platforms mitigate the food safety data analysis, performing services such as data mining, data visualization, predictive analytics for food safety, and statistical investigation.
- The *application layer* is responsible for generating user-friendly applications to access specific portions of the collected information in real-time at the touch of fingertips, including possible food safety violations. These applications present the management with dashboards on food location, status and environment-based conditions of food. Management will also have access to other applications necessary for preparing and delivering safe food to hospital patients.

10.4.2 Advantages of adopting IoT-based wearable devices in healthcare food preparation

Adopting digital IoT-based wearable devices can ensure food safety as well as offer tangible benefits to those responsible for food preparation in healthcare institutions. Implementation of these wearable devices will provide support to the persons wearing them to assure patient safety and improve food operations at the same time. Four key advantages have been identified:

1. *Transparency*: Utilizing IoT-based wearable devices can ensure food safety by electronically displaying alerts or alarms for the concerned employees, indicating when checks are due, reporting the results to the appropriate management personnel, and transmitting data to the cloud storage. Once data is organized and secure in the cloud storage, any authorized stakeholder or hospital management personnel can view the information in real-time and take immediate necessary actions to address any issues. This advantage reduces the delay in addressing critical food safety matters and eliminates the need to sift through piles of paperwork to spot problems.
2. *Safety*: Most food preparation checklists in healthcare facilities still follow paper-based systems and are often inefficient due to a lack of guidance, control, and authenticity. Paper-based systems do not assure management that

all of the food safety procedures are being followed. The paperwork is often filled out at staff convenience instead of requested times, information is prone to fabrication, and there is no exact time record of when and by whom the check is carried out. Digital IoT-linked wearable food safety devices can address these issues effectively and provide a robust method to record exactly when a check was carried out and which member of staff was responsible. The information collected by the system can instantly be shared with all the stakeholders, and this ultimately reduces the time required for inspections and routine reporting.

3. *Productivity*: Labor costs in the food preparation environment are often high; therefore, enhancing work environment productivity and effectiveness without compromising food safety is of utter importance. Filling out paperwork related to food safety and operations takes a considerable amount of employee time, and by swapping to IoT-based digital systems, employees can be freed up to do other preparation activities. For example, in an IoT environment, the temperature sensors placed within chillers and freezers can replace paper-based systems and automate the process of recording temperature readings without human interference.
4. *Efficiency*: Because of the high staff turnover in many food service jobs, it must be ensured that all staff follow GMPs in the food environment. Conventional paper-based system often do not describe the correct procedures to prepare safe food. IoT-based wearable systems can support staff through all the steps of food safety checks and even provide corrective actions if some parameters do not fall within the set range, ensuring safe food is delivered to patients and their visitors.

10.4.3 Utilizing wearable IoT technologies to solve food safety issues in healthcare preparation

As mentioned before, hospital preparation systems follow the HACCP system to ensure that food safety is maintained throughout the whole process of receiving and preparing food. HACCP system identified 11 CCPs within a hospital preparation operation that could lead to hazards. These issues are currently managed through paper-based systems, and IoT-based technologies can be successfully adopted to replace them. Table 10.5 shows the CCPs, hazards, control measures to prevent hazards, quality checks required, and paper-based records maintained by hospital preparation staff that can addressed by the CCP. Fig. 10.4 gives a visual representation of IoT devices that can be used to monitor the food safety in real time. Wearable technology such as smart glasses and barcode scanner gloves can be easily used with food production, transport, and storage systems through which management can keep track of inventory and manage temperature control throughout the supply chain [25]. Fortunately, NSF International and Google are collaborating to develop an application of Google glasses to utilize them for food

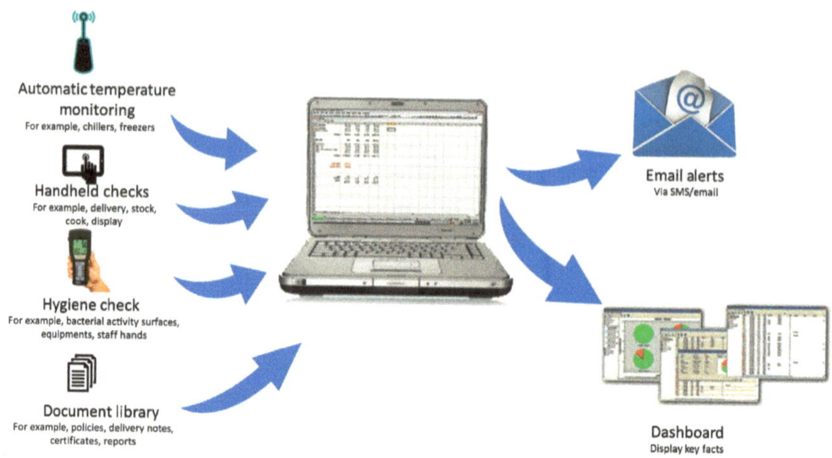

FIGURE 10.4

Process suitable for automation using IoT.

safety auditing and training [26]. Temperature sensors, handheld devices such as smartphones, and hygiene meters can collect data and transfer it to the cloud storage where it can be analyzed, and performance reports or evidence of any breaches to food safety can be sent to hospital management. Table 10.4 lists the wearable IoT-based devices currently being used or that could be used in the future to manage food operations within the healthcare sector to ensure the delivery of safe food.

Table 10.5 lists the paper-based logs that could be replaced by IoT technology currently available in the market and what level of automation can be achieved. It illustrates which processes can be successfully automated or partially automated (human assistance needed) through the implementation of IoT technology.

10.4.4 Features and requirements of sensors used in wearable devices for food safety applications

Several sensors are readily available in the market that can be used in wearable devices to improve the food safety depending on the desired results. Sensors are vital components of wearable devices; this section describes the features and requirements of sensors required in food safety application.

Environmental sensors: Environmental sensors are used for measuring, monitoring, and recording environmental parameters [27] such as measuring temperature, humidity, heat losses, and detecting food spoilage using gas sensors. These sensors are crucial to ensure that food is cooked or stored at right temperatures and to be sure that the food is fit for human consumption.

Table 10.4 Wearable technologies with their capabilities and food safety applications.

Type of wearable technology	Capabilities	Food safety applications
Smart Staff ID card	Activity tracking	Authentication, data entry crucial for food safety
Smart Eyewear	Visualization & communication	Stock control, training, education, temperature recording
Smart Clothing	Location tracking	Food handling
Wearable Camera	Photographic or audio-video evidence	To keep the record of food processes and any food abuse
Smart Swabs	Real-time bacterial activity	Food, personal, and equipment hygiene

Table 10.5 CCPs, hazards, control measures, checks, and records.

Process	Hazards	Control measures	Checks	Records
Purchasing CCP 1	MCP	Products to be purchased from authorized approved suppliers; approved suppliers list to be recorded; suppliers to deliver high-risk and perishable products at a maximum temperature of 0°C–3°C and frozen foods at a maximum temperature of 18°C; performance of each supplier will be reviewed, and any nonsatisfactory performance will lead to revoking of contract; the temperature of products to be recorded on delivery notes	Goods historical records; auditing of supplier's premises	Suppliers list available with preparation management

(Continued)

Table 10.5 CCPs, hazards, control measures, checks, and records. *Continued*

Process	Hazards	Control measures	Checks	Records
Receipt CCP2	MCP	Goods received and handled in appropriate manner to avoid the risk of damage and contamination; nonfood items not to be accepted during food deliveries to avoid contamination; good-in staff follow hygiene code of institution; enforcement of the maximum temperature allowed for unloading food items (frozen target 18°C and chilled target 0°C–5°C); inspected goods only transferred to storage; noncompliant goods returned to supplier or destroyed; microbial testing of produce	Visual check for appearance, quality, packaging condition, any signs of infestation; measurement of food temperatures; visual check of delivery vehicles	Goods records, laboratory results, product rejection/ segregation, delivery notes
Storage CCP3	MCP	Goods stored at suitable temperatures (frozen 18°C, chilled—0°C–3°C); storing area needs to be clean, hygienic, pest-free; prevent cross-contamination; staff mandated to follow personal hygienic code of institution; stock rotation FIFO, plan for suitable action to be taken in case of nonconformance; frozen food, once defrosted, will be dated as per manufacturer guidelines; in the event of overproduction cooked items will be frozen, dated, and labeled	Daily refrigeration checks; weekly food storage checks; hygiene checks; pest inspection	Temperature records, hygiene audit; pest inspection records

(*Continued*)

Table 10.5 CCPs, hazards, control measures, checks, and records. *Continued*

Process	Hazards	Control measures	Checks	Records
Preparation CCP4	MCP	Food types – raw and cooked prepared in their designated areas and equipment; minimize the time that high-risk food kept at room temperature; food is covered and protected to avoid physical and chemical contamination and stored in appropriate locations; surfaces and equipment are sanitized after use; staff follow food hygiene code; all production areas are kept clean, hygienic, and pest-free; corrective action to be taken in case of noncompliance; new staff to be fully trained; blast chilling and refrigeration procedure to be followed	Daily cleaning check program; reporting of noncompliance	Pest control records; daily hygiene reports; staff training
Cooking CCP5	MCP	Raw food to be cooked at 75°C for a minimum of 2 min; food to be served immediately; food for storage to be protected from contamination by storing in the right location; cooked utensils to be washed and sanitized after use; staff to follow hygiene rules; staff to be fully trained for food handling activities	Kitchen and equipment cleaning program; measurement of temperature during cooking using probe thermometer; measurement of cooking time	Cooking check, staff training; staff appearance; noncompliance reports

(Continued)

Table 10.5 CCPs, hazards, control measures, checks, and records. *Continued*

Process	Hazards	Control measures	Checks	Records
Chilling/ Blast Chill CCP6	MCP	Blast chilling will commence within 30 min following the completion of the cooking process; food will be blast chilled to a maximum core temperature of +5°C (critical limit +8°C) without undue delay and within 90 min; all chilled food will be stored at below +5°C	Measurement of temperature during chilling using sanitized probe thermometer; measurement of chilling time.	Chilling check and blast chilling checks; quantity check; staff training; staff appearance; noncompliance reports
Food Service CCP7	MCP	Only food that already exceeds a temperature of +80°C (critical limit +70°C) may be placed on plating up system, or hot hold, or displayed hot; chilled foods will be displayed at a maximum temperature of +5°C (critical limit +8°C)	Two temperature checks during service on hold/ display units at random times; temperature checks of selected high-risk foods	Temperature check on high-risk foods; temperature; staff training; staff appearance; noncompliance reports
Reheating CCP8	MCP	Food will be reheated to a minimum core temperature of +75°C (critical limit +70°C) for a minimum period of 2 min	Measurement of temperature during reheating using a sanitized probe thermometer; control of reheating	Reheating check; staff training; staff appearance, noncompliance reports
Trolley Service CCP9	MCP	Hot foods will be held and served at a minimum core temperature of +80°C (critical limit +70°C); chilled foods will be held and served at a maximum temperature of +5°C (critical limit +8°C)	Check delivery carts at random times; temperature checks of selected high-risk foods at random times	Ward service; temperature check of selected high-risk foods; staff training, staff appearance, noncompliance reports

(*Continued*)

Table 10.5 CCPs, hazards, control measures, checks, and records.
Continued

Process	Hazards	Control measures	Checks	Records
Ward service CCP10	MCP	Hot foods will be held and served at a minimum temperature of +70°C; chilled foods will be held and served at a maximum temperature of +5°C (critical limit +8°C)		Every service temperature checks of selected high-risk foods at random times
Calibration CCP11	MCP	Master probe will be checked every 3 months using the calibration testing caps at 0°C and 100°C; the calibration testing caps will be sent for calibration every 12 months and have a current calibration test certificate on file; all probes will be calibrated using boiling water at 100°C, and also ice at 0°C; at the time of calibration, all temperature probes outside the tolerance of +2.0°C or 2.0°C will be removed from use		Calibration records will be updated on a monthly basis

Biosensors: Biosensors are used to analyze substances by converting a biological response into a signal [28]. Bacterial activity in food processing or handling areas and heavy metal contamination in food can be successfully detected using biosensors [29]. It is an easy and efficient way to determine whether food is contaminated with harmful pathogens such as *E. coli*, *salmonella*, etc., or to find out if the hygiene process was effective.

Positioning and tracking sensors: Positioning or tracking sensors such as GPS are the most widely used sensors on wearable devices or equipment to track the location and movement [30] of food handlers and food stocks. It can be used to identify if the food had undergone all of the necessary processes such as cooking or cooling before served to patients.

Miscellaneous sensors: These consist of other sensors freely available on the market. Body cameras and smart glasses are the sensors used for visual recording or evidence. Other communication sensor modules such as Bluetooth, RFID, and WiFi provide communication and data exchange functioning to wearable devices [31]. These are used for location tracking of staff and food trolleys and to deliver the right food to the right patients.

10.5 Discussion

Providing quality and safe food to the patients as a means of helping them recover faster is the utmost priority. Through various IoT-based wearable sensing devices discussed above, it is now possible to detect pathogens, contaminants, and gases from unfit food, and to get alerts on changes in the environmental conditions of food at any point between receiving food and serving to the patients. Such a sensing system can ensure quality control with patients being served with fresh, safe, and nutritious food products and at the same time improve the response time of the preparation and management staff in case of a breach in food safety. This novel concept has a significant amount of benefits when compared to the traditional paper-based system currently followed in the hospital food preparation sector for monitoring food safety. Despite its numerous offerings to improve food quality standards, the adoption of IoT-based devices in the hospital food preparation sector is slower than expected. However, it is on the rise due to exciting possibilities for food regulatory authorities. Obviously, there are hurdles that need to be overcome to commercialize this technology effectively. But continuous research and developments in IoT-based wearable sensing technology will help in its adoption and ensure that in the future it plays an important role in making the food safer, healthier, and more nutritious.

10.6 Conclusion

This chapter highlighted the challenges faced by the healthcare sector to overcome the food safety issues and the actions needed to eliminate potential food hazards that arise in the process of food preparation. With the primary aim of ensuring the preparation system is providing safe food, proactive hospital management focuses on preventing the food hazards rather than reacting to food safety issues after they occur. The scale of food safety breaches in recent times have compelled food providers and authorities to realize that new strategies are needed to counter these challenges. Robust implementation of HACCP processes is needed to ensure food safety across the entire food supply chain and bring the procedural change needed to food safety issues. To monitor and track every movement of food in real time, the IoT devices such as mobiles, handheld devices, and other wearable technologies are utilized. This provides flexibility and can be deployed at various critical food safety hotspots within the preparation system: from the point where food ingredients are received to the point food reaches the consumer (patients, visitors, etc.). These IoT-based devices prompt the actions expected of the preparation staff and records the precise time-stamped

Table 10.6 Automation achieved through implementation of IoT

Paper-based logs	Reasons	IoT technology	Automation level
Visitor Health Declaration	To ensure any illness symptoms (vomiting/diarrhea, etc.) or foreign objects (jewelry, pen, etc.); compliance with personal protective equipment (hair covering, safety shoes, etc.)	Touch screen at reception	Partially
Supplier Assessment Form	To ensure supplier is following all quality checks on ingredients before delivering it to the hospital	An electronic copy can be maintained	Partially
Daily Delivery Monitoring Form	Addresses CCP1 & CCP2 (crucial data supplier name, time, chilled/ frozen, use by date, temperature, accept/reject)	Handheld device with probe	Partially
Refrigerator/ Freezer Temperature Monitoring (in all departments)	To record foods are stored at the right temperature; checks are performed every hour	RFID temp probe	Fully
Food Production Sheet	To record start/finish time for a process (cooking/chilling/reheating/ holding)	Handheld device	Partially
Food Service Staff Feeding Monitoring Form	To record food temperature at service	RFID temp probe	Fully
Chiller Record Sheet	To record the time and temperature when a product is sent for chilling or removed	Handheld device with probe	Partially
Personal Hygiene Record	To record if employees are adhering to hygiene rules (correct Personal Protective Equipment (PPE), jewelry, nail varnish, etc.)	Handheld device with camera	Partially
Cleaning Schedules	Daily cleaning (food processing equipment, floors, surfaces, etc.)	Handheld ATP hygiene meter	Partially
Calibration of temperature probes	To ensure if temperature probes are showing the correct readings	Handheld device with master probe	Partially

audit trail of its completion. This is a major development in eliminating food safety issues in the health sector preparation system where patients are vulnerable to unsafe food. The benefits over conventional paper-based logs are tremendous, including the biggest saving in the cost of maintaining stringent HACCP compliance requirements. Handheld or wearable IoT solutions support the preparation staff by adhering to laws and processes and ensuring the food consumed is safer (Table 10.6).

References

[1] ITV News, Dirty hospitals serving out-of-date food are putting patients at risk of food poisoning, investigation finds. Available from: http://www.itv.com/news/2016-11-02/patients-at-risk-of-food-poisoning-as-dirty-hospitals-are-serving-out-of-date-food-investigation-finds/, 2016 (accessed 23.02.18).

[2] R. Garayoa, C. Abundancia, M. Diez-Leturia, A. Vitas, Essential tools for food safety surveillance in preparation services: on-site inspections and control of high risk cross-contamination surfaces, Food. Control. 75 (2017) 48–54.

[3] A. Jones, et al., Estimating the prevalence of food risk increasing behaviours in UK kitchens, PLoS One 12 (6) (2017).

[4] C. Bhatt, N. Dey, A. Ashour (Eds.), Internet of Things and Big Data Technologies for Next Generation Healthcare, first ed., Springer, Cham, 2017.

[5] L. Atzori, A. Iera, G. Morabito, The internet of things: a survey, Comput. Networks 54 (15) (2010) 2787–2805.

[6] Y. Zhang, T. Qu, O. Ho, G. Huang, Real-time work-in-progress management for smart object-enabled ubiquitous shop-floor environment, Int. J. Comput. Integr. Manuf 24 (5) (2011) 431–445.

[7] Y. Zhang, et al., Real-time information capturing and integration framework of the internet of manufacturing things, Int. J. Comput. Integr. Manuf. 3052 (2014) 1–12.

[8] W. Zhou, S. Piramuthu. IoT and supply chain traceability, in: 1st International Conference on Future Network Systems and Security, Paris, 2015.

[9] T. Bosona, G. Gebresenbet, Food traceability as an integral part of logistics management in food and agricultural supply chain, Food. Control. 33 (1) (2013) 32–48.

[10] Y. Zhang, J. Wang, Y. Liu, Game theory based real-time multi-objective flexible job shop scheduling considering environmental impact, J. Clean. Prod. 167 (2017) 665–679.

[11] Y. Zhang, et al., An optimization method for shopfloor material handling based on real-time and multi-source manufacturing data, Int. J. Prod. Econ. 165 (2015) 282–292.

[12] Y. Zhang, D. Xi, R. Li, S. Sun, Task-driven manufacturing cloud service proactive discovery and optimal configuration method, Int. J. Adv. Manuf. Technol. 84 (1–4) (2016) 29–45.

[13] D. Barrie, The provision of food and preparation services in hospital, J. Hosp. Infect. 33 (1996) 13–33.

[14] WHO, World Health Organisation. Available from: http://www.who.int/foodsafety/areas_work/food-hygiene/en/, n.d. (accessed 28.03.18).

[15] NHS, Food poisoning. Available from: https://www.nhs.uk/conditions/food-poisoning/, 2017 (accessed 25.03.18).

[16] B. Ray, A. Bhunia, Fundamental Food Microbiology, fourth ed., CRC Press, 2008.

[17] NHS, Allergies. Available from: https://www.nhs.uk/conditions/allergies/symptoms/, 2016 (accessed 25.03.18).

[18] A. Baltazar, J. Figueiredo, A. Ferreira, 2017. Safety control evaluation of food storage in mass caterer, in: Occupational Safety and Hygiene V, Volume Proceedings of SHO 2017, pp. 525–528.
[19] S. Mortimore, C. Wallace, HACCP A Practical Approach, third ed., Springer, US, 2013.
[20] B.M. Lund, S.J. O'Brien, Microbiological safety of food in hospitals and other healthcare settings, J. Hosp. Infect. 73 (2009) 109–120.
[21] B.M. Lund, Microbiological food safety for vulnerable people, Int. J. Environ. Res. Public. Health 12 (2015) 10117–10132.
[22] GOV.UK, Good manufacturing practice and good distribution practice. Available from: https://www.gov.uk/guidance/good-manufacturing-practice-and-good-distribution-practice, 2017 (accessed 29.03.18).
[23] R.H. Schmidt, G.E. Rodrick, Food Safety Handbook, John Wiley & Sons, 2005.
[24] M. Rossi, E. Stedefeldt, D. da Cunha, V. de Russo, Food safety knowledge, optimistic bias and risk perception among food handlers in institutional food services, Food. Control. 73 (2017) 681–688.
[25] H. Smith, Wearable tech for good warehouse logistics. Available from: http://foodindustryexecutive.com/2017/12/wearable-tech-food-warehouse-logistics/, 2018.
[26] R. Chester, The future of food safety: the revolution is on our doorsteps. Available from: https://www.newfoodmagazine.com/article/41390/future-food-safety-revolution-doorsteps/, 2018.
[27] K. Jones, Environmental sensors, in: H. Meixner, R. Jones (Eds.), Sensors: Micro- and Nanosensor Technology-Trends in Sensor Markets, 1995, VCH Verlagsgesellschaft mbH, Weinheim, Germany, 2008, pp. 451–489.
[28] V. Velusamy, et al., An overview of food-borne pathogen detection: In the perspective of biosensors, Biotechnol. Adv. 28 (2) (2010) 232–254.
[29] J. Yasmin, M. Ahmed, B. Cho, Biosensors and their applications in food safety: a review, J. Biosyst. Eng. 41 (3) (2016) 240–254.
[30] H. Kung, L. Zhang, X. Xie, W. Ma, Mining interesting locations and travel sequences from GPS trajectories. in: 18th International Conference on World Wide Web – WWW'09, Madrid, Spain, 2009.
[31] J. Baek, et al., BBUNS: bluetooth beacon-based underground navigation system to support mine haulage operations, Minerals 7 (228) (2017).

Further reading

GOV.UK, Food safety – your responsibilities. Available from: https://www.gov.uk/food-safety-your-responsibilities/food-hygiene, n.d. (accessed 04.03.18).

Index

Note: Page numbers followed by "*f*" and "*t*" refer to figures and tables, respectively.

T

U

V

W

X

Z

FSC
www.fsc.org
MIX
Papier | Fördert gute Waldnutzung
FSC® C083411

Kolibri 360 GmbH
Ferdinand-Jühlke-Straße 7
99095 Erfurt, Deutschland
produktsicherheit@kolibri360.de